Soil Contamination - Recent Advances and Future Perspectives

Edited by Adnan Mustafa and Muhammad Naveed

Published in London, United Kingdom

Soil Contamination - Recent Advances and Future Perspectives
http://dx.doi.org/10.5772/intechopen.100944
Edited by Adnan Mustafa and Muhammad Naveed

Contributors
Adnan Mustafa, Ajayi Ochechevesho Joan, Anahí C. Meza García, Anais Venisseau, Clarisse Balland-Bolou-Bi, Claudia Coelho, Dorota Pikuła, Emaan Amjad, Evangeline Njanja, Fabiola Tonleu Dontsi, Fahmeeda Naheed, Florence Brondeau, Gullit Deffo, Hassan Ali, Igbonomi Emmanuel Sunday, Ignas Tonle Kenfack, Jean-Philippe Bedell, José G. Guadalupe Chan Quijano, Kevin Yemele Tajeu, Kobra Jalali, Lionnel Averie Vomo, Muhammad Fraz Ali, Muhammad Naveed, My Dung Jusselme, Olivier Roques, Philippe Marchand, Qudsia Saeed, Ranil Clément Temgoua Tonleu, Razieh Lamoochi, Saddam Hussain, Sadia Baloch, Shaharyar Khosa, Shahzaib Ali, Sherman Lesly Zambou Jiokeng, Sidra Sohail, Usman Zulfiqar, Yves Perrodin

First published in London, United Kingdom, 2024 by IntechOpen
IntechOpen is the global imprint of INTECHOPEN LIMITED, registered in England and Wales, registration number: 11086078, 167-169 Great Portland Street, London, W1W 5PF, United Kingdom

British Library Cataloguing-in-Publication Data
A catalogue record for this book is available from the British Library

Additional hard and PDF copies can be obtained from orders@intechopen.com

Soil Contamination - Recent Advances and Future Perspectives
Edited by Adnan Mustafa and Muhammad Naveed
p. cm.

This title is part of the Environmental Sciences Book Series, Volume 17
Topic: Pollution
Series Editor: J. Kevin Summers
Topic Editors: Ismail Md. Mofizur Rahman and Zinnat Ara Begum

Print ISBN 978-1-80356-827-0
Online ISBN 978-1-80356-828-7
eBook (PDF) ISBN 978-1-80356-829-4
ISSN 2754-6713

For EU product safety concerns:
IN TECH d.o.o., Prolaz Marije Krucifikse Kozulić 3, 51000 Rijeka, Croatia,
info@intechopen.com or visit our website at intechopen.com.

IntechOpen Book Series

Environmental Sciences

Volume 17

Aims and Scope of the Series

Scientists have long researched to understand the environment and man's place in it. The search for this knowledge grows in importance as rapid increases in population and economic development intensify humans' stresses on ecosystems. Fortunately, rapid increases in multiple scientific areas are advancing our understanding of environmental sciences. Breakthroughs in computing, molecular biology, ecology, and sustainability science are enhancing our ability to utilize environmental sciences to address real-world problems.

The four topics of this book series - Pollution; Environmental Resilience and Management; Ecosystems and Biodiversity; and Water Science - will address important areas of advancement in the environmental sciences. They will represent an excellent initial grouping of published works on these critical topics.

Meet the Series Editor

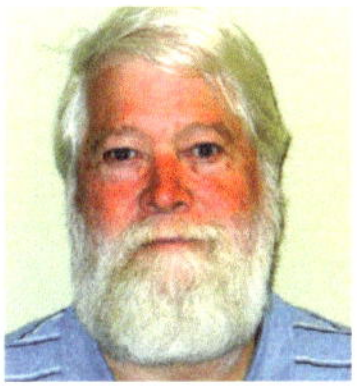

J. Kevin Summers is a Senior Research Ecologist at the Environmental Protection Agency's (EPA) Gulf Ecosystem Measurement and Modeling Division. He is currently working with colleagues in the Sustainable and Healthy Communities Program to develop an index of community resilience to natural hazards, an index of human well-being that can be linked to changes in the ecosystem, social and economic services, and a community sustainability tool for communities with populations under 40,000. He leads research efforts for indicator and indices development. Dr. Summers is a systems ecologist and began his career at the EPA in 1989 and has worked in various programs and capacities. This includes leading the National Coastal Assessment in collaboration with the Office of Water which culminated in the award-winning National Coastal Condition Report series (four volumes between 2001 and 2012), and which integrates water quality, sediment quality, habitat, and biological data to assess the ecosystem condition of the United States estuaries. He was acting National Program Director for Ecology for the EPA between 2004 and 2006. He has authored approximately 150 peer-reviewed journal articles, book chapters, and reports and has received many awards for technical accomplishments from the EPA and from outside of the agency. Dr. Summers holds a BA in Zoology and Psychology, an MA in Ecology, and Ph.D. in Systems Ecology/Biology.

Meet the Volume Editors

Dr. Adnan Mustafa completed his Ph.D. in Soil Science at the Graduate School of the Chinese Academy of Agricultural Sciences, Beijing, China. He worked as a Postdoctoral Researcher at the Biology Centre, Czech Academy of Sciences and as an Assistant Professor at the Institute of Chemistry and Technology of Environmental Protection, Brno University of Technology, Czech Republic. He is currently working as a Research Fellow at the South China Botanical Garden, Chinese Academy of Sciences, Guangzhou, China. His research focuses on soil organic carbon (SOC) accumulation mechanisms, plant-microbe interactions, and biochar production and utilization for agricultural crop production and environmental remediation. He is actively involved in the bioremediation of contaminated soils using organic and inorganic amendments in addition to exploiting plant-microbe interactions. He has published more than 100 refereed journal articles, many of which sought to explore the effectiveness of innovative soil amendments and plant growth-promoting rhizobacteria (PGPR) for improving crop performance and soil resilience under various abiotic stresses.

Dr. Muhammad Naveed is an Associate Professor of Soil Science at the Institute of Soil and Environmental Sciences, University of Agriculture, Pakistan. He received his Ph.D. in Soil Microbiology/Ecology from the University of Natural Resources and Life Sciences, Austria. He has received many national and international awards, including the Salimuzzaman Siddiqui Prize for Applied Science - Technology in Agricultural Sciences from Pakistan and the Gregor Mendel Society Award for Visionary Ideas on Genetics from Austria. Dr. Naveed's research focuses on the physiological and molecular mechanisms of biotic and abiotic stress tolerance in plants. He is actively working on plant-microbe interactions under stress conditions. He has published more than 130 research articles in peer-reviewed journals and more than 15 book chapters. He also has three patents to his credit.

Contents

Preface

This edited volume is a collection of reviewed and relevant research chapters concerning developments in soil contamination. It includes scholarly contributions written by various authors and edited by a group of experts in the field. It is designed for scholars and specialists in the field.

The book contains nine chapters. Chapter 1, "Biochar Application for Soil Quality Improvement: An Overview", reviews the potential of biochar to enhance soil quality attributes, particularly those pertaining to soil's physical, chemical, and biological properties. Chapter 2, "Perspective Chapter: Sample Preparation Techniques for Electrochemical Analysis of Pesticides and Heavy Metals in Environmental and Food Samples", describes the steps involved in preparing samples for electrochemical analysis. Chapter 3, "Influence of Sewage Sludge and Heavy Fertilization on Nitrate Leaching in Soils: An Overview", focuses on various factors that affect nitrate leaching in soil, including soil texture, rate of fertilization, season, and climate. It also presents strategies to reduce nitrate leaching and examines the nitrogen conversion mechanism in sewage sludge. Chapter 4, "Occurrence, Bioaccumulation and Effects of Legacy and Emerging Brominated Retardants in Earthworms", deals with the transfer processes of emerging brominated flame retardants in earthworms, as well as the potential trophic biomagnification. Chapter 5, "Ecological Risk Index, Implications, and its Bioconcentration Ratio in Soils Contaminated with Heavy Metals", focuses on the implications and consequences that heavy metals cause. Chapter 6, "Effect of the Degree of Soil Contamination with Cd, Zn, Cu and Zn on Its Content in the Forder Crops and Mobility in the Soil Profile", describes the correlation between the amount of leached metals and the level of soil contamination. Chapter 7, "Investigating Bacterial Communities Resistant to Heavy Metal and PAHs Pollutants in the Persian Gulf and Their Ability to Remove These Compounds (a Review)", reviews studies conducted on isolating and identifying native bacteria and evaluating their ability to remove heavy metals and polycyclic aromatic hydrocarbons in the Persian Gulf. Chapter 8, "Bacterial Assisted Phytoremediation", explores the interaction between bacteria, plants, and various pollutants in various ecosystems. Chapter 9, "Perspective Chapter: Can Natural Attenuation be Considered as an Effective Solution for Soil Remediation?" discusses the principle of this technology, the role of the living actors and the interactions between plant, fauna, and microorganisms, the advantages and disadvantages of the technique, and the efficiency of this technique in comparison with other techniques such as phytoremediation and bioremediation.

IntechOpen Team

Chapter 1

Biochar Application for Soil Quality Improvement: An Overview

Hassan Ali, Shahzaib Ali, Sadia Baloch, Fahmeeda Naheed, Emaan Amjad, Qudsia Saeed, Muhammad Naveed and Adnan Mustafa

Abstract

Soil as a renewable resource has a key role to play in sustainable crop production, soil management, and combating food insecurity. The overapplication of fertilizers in this regard has resulted in decreased soil health and productivity. Biochar application in this respect has received increasing attention of the scientific community due to its role in soil quality improvement. This is especially true in the face of global climate change and to the nature of biochar being a carbon (C)-rich compound. In this chapter, the potential of biochar to enhance soil quality attributes, particularly those pertaining to soil's physical, chemical, and biological properties, is comprehensively reviewed. Special attention is directed toward the distinctive properties of biochars sourced from various feedstocks, elucidating their subsequent effects on soil quality. This sheds light on potential directions for future studies in this field.

Keywords: biochar, soil health, sustainable environment, soil improvement, soil quality

1. Introduction

Climate change negatively impacts crop growth. Salinized, weak alkaline soils; temperature fluctuations; and latitudinal factors require urgent scientific attention [1, 2]. Soil, a complex ecosystem, plays a vital role in agriculture, ecosystem services, climate change mitigation, landscape protection, and human development [3, 4]. Long-term cultivation can lead to degradation, resulting in reduced soil organic matter, severe erosion, and diminished aggregate stability [5, 6]. Soil quality directly impacts ecosystems, affecting crop production, biota, and human well-being, contributing to a healthy environment [7, 8]. Various factors, including pollution levels, bacterial activity, climate, and land use, can alter soil characteristics [9]. Green approaches, like soil amendment, offer effective means to reduce salinity by enhancing physical and chemical properties and nutrient distribution. These approaches also promote microorganism growth and restore soil health [10–13]. The application of additives is a promising method to enhance soil characteristics. In recent years, biochar (B) has gained significant interest as a versatile soil fertilizer, owing to its high porosity, abundant surface functional groups, strong adsorption capacity, and recalcitrant carbon content [14].

IntechOpen

Biochar (BC) is plant-derived charcoal used to absorb carbon dioxide when incorporated into soil [15, 16]. It plays a crucial role in long-term climate change mitigation [17] and enhances soil biochemical properties [18]. BC also facilitates large-scale disposal of waste biomass [19]. Studies highlight its significance as a carbon source and soil enhancer, sequestering pollutants and enhancing fertility [20]. Its elemental composition includes essential elements like C, N, and H, alongside lower-nutrient elements [21]. Adding BC improves soil properties such as porosity, bulk density, and water availability [22]. Pyrolysis releases volatile compounds, increasing BC's surface area and creating pores akin to honeycombs, enhancing water and nutrient retention [23, 24].

BC incorporation enhances soil structure, aeration, nutrient availability, and alters microbial communities, fostering crop growth and yield [25–27]. It raises soil pH and enhances cation exchange capacity [28, 29]. Amended soils exhibit heightened fertility and nutrient retention [30], with significantly boosted extractable nutrient levels [31]. BC increases hydraulic conductivity, base saturation, and nutrient levels and reduces erosion rate [32, 33]. It halts the leaching of nitrate nitrogen from soils [34]. The diverse pore sizes of BC influence microorganism habitats, promoting beneficial bacterial survival and enhancing bacterial structure [35–41]. BC's effects on soil microorganisms are influenced by nutrient availability, microbial community composition, plant–microbe signaling, and habitat creation [42].

Climate change disrupts the vital carbon storage in soil, impacting global food production and greenhouse gas emissions [43–46]. Methane (CH_4) and nitrous oxide (N_2O), potent greenhouse gases, primarily from agricultural land, have a significant warming effect [47]. Biochar is pivotal in curbing emissions and averting climate change by enhancing soil quality, reducing pollution, and stabilizing carbon [48–50]. It is technically and economically feasible for carbon capture and storage [51], disrupting the carbon cycle and preventing emissions from reentering the atmosphere [52, 53]. This fosters soil health, lowers emissions, and boosts carbon sequestration [54–56].

2. Literature review methodology

A set of thematic keywords were used for exploring the relevant research articles and review papers, such as biochar, feedstock, soil quality, soil fertility, soil pH, soil physicochemical properties, biochar amendments, soil organic matter, carbon sequestration, the role of biochar amendment, plant growth and yield attributes, crop performance, and so forth. In this regard, five different databases were used, namely, Research Gate, Google Scholar, Science Direct, Web of Science, and Scopus. These databases comprise extensively used huge collections of related mainstream research articles. Following this, multiple research and review articles were selected within the era 2000–2022. Furthermore, related research articles referenced in the abovementioned papers were also reviewed.

3. Application of biochar in soil quality improvement

Biochar is considered as a potential soil amendment, as its addition improves the C accumulation (60–80%) in the soil, which is central to the enhancement in the soil properties, and hence, biochar technology could be very effective for soils having low organic matter contents (OM) [22]. The following sections discuss the role of biochar as a potential amendment to enhance soil quality.

3.1 Improvement in soil bulk density

Soil bulk density, a critical indicator of soil physical properties, is closely linked to soil tightness [57]. It reflects soil compaction and health and influences crucial aspects of the plant life cycle [58]. Enhanced soil porosity and aeration, facilitated by biochar, lead to reduced bulk density [59]. Biochar's lower bulk density contributes to decreased soil bulk density and improved water holding capacity due to its larger surface area [60–62]. Numerous studies have reported a significant negative impact of biochar addition on soil bulk density [63, 64]. Głąb et al. [65] found a 35% reduction in bulk density with the incorporation of 4% biochar. Similarly, Qin et al. [66] documented a decrease in bulk density with an increase in overall soil porosity after introducing biochar. The influence of biochar on soil bulk density is closely tied to soil texture; coarser-textured soils experience more pronounced reductions compared to finer-textured soils [67]. This may be attributed to the higher bulk density of coarse-textured soils (~1.6 g cm^{-3}) compared to biochar (~ 0.6 g cm^{-3}), allowing interactions between the biochar and soil particles, resulting in a decline in the final bulk density of the soil. Additionally, biochar's high porosity and sandy soils' lower porosity contribute to decreased bulk density [68]. Furthermore, the nature of feedstock material and pyrolysis temperature significantly influence bulk density, with higher pyrolysis temperatures leading to lower bulk density [69]. The rate of biochar application also affects bulk density, generally showing a negative correlation with application rate [58].

3.2 Improvement in surface area and soil porosity

Biochar significantly influences soil porosity through its pore distribution, particle size, and connectivity [57]. Studies have reported an increase in soil porosity following biochar addition, ranging from 14% to 64% [22]. Higher pyrolysis temperatures enhance soil porosity, potentially due to organic matter decomposition and the formation of micropores [70, 71]. The impact of biochar on soil porosity depends on soil texture and type, with coarse-textured soils showing a greater increase compared to fine-textured soils, likely due to their lower inherent porosity [72]. In sandy soils, biochar improves porosity by enhancing mechanical interactions with soil particles, leading to notable improvements [73]. Increased porosity positively affects soil microbial activities, and different pore size distributions may influence specific microbial species [74]. Additionally, Seyedsadr et al. [75] found that biochar addition increased total porosity and reduced bulk density compared to compost. They suggested several mechanisms for improved soil porosity, including direct pore effects within biochar, accommodation pores formed between soil aggregates and biochar, and enhanced persistence of soil pores through increased aggregate stability [76].

3.3 Improvement in soil aggregate stability

Soil aggregate stability is crucial for resisting mechanical stresses like surface runoff, water erosion, and precipitation effects [58, 77]. Disintegration of aggregates leads to fine particles susceptible to erosion, potentially forming a soil crust upon re-sedimentation [78]. Biochar application significantly impacts aggregate stability through interactions with soil mineral surfaces, facilitated by oxidized carboxylic groups on biochar particles [79]. This interaction fosters aggregate formation, creating stable microaggregates and larger pores, ultimately enhancing stability and structure [80]. Studies have reported an increase in soil aggregate stability ranging from 6 to 217% after biochar addition [32, 81]. This improvement may be attributed to the high carbon content of biochar, forming bonds with various oxides, and organic matter providing a food source for microorganisms. This leads to increased microbial activity and the secretion of mucilage, contributing to stable aggregates [82]. Additionally, biochar may enhance aggregate resistance against clay swelling and slaking [83].

3.4 Improvement in soil consistency

Soil consistency attributes, such as plastic limit, liquid limit, and plasticity index, are crucial for designing stable slope systems [84]. Understanding soil consistency is vital for engineering applications and agronomic contexts, including tillage operations and compaction. Additionally, biochar application can enhance soil consistency by increasing organic carbon concentrations [85]. Different types of biochar applied to sandy soils at various rates resulted in significant improvements in plasticity index (48–99%) and liquid limits (8–22%) [85]. This improvement is primarily attributed to biochar's larger surface area and increased porosity [86, 87]. Laboratory experiments by Choudhary et al. [88] showed that biochar addition improved soil consistency limits up to a specific weight limit, leading to enhanced physical properties of the soil.

3.5 Improvement in cation exchange capacity (CEC)

The CEC indicates the ability of soil to absorb, retain, and exchange cations, and enhancing the number of exchange sites for cations can augment the soil CEC contents. Soils having a high CEC can readily absorb K^+, NH_4^+, Mg^{2+}, and Ca^{2+} and improve the utilization of nutrient ions in soils [89]. The CEC is largely pH-dependent particularly on tropical soils, and some biochars can raise the pH of the soil as well as soil CEC [90–92]. Use of peanut shell-derived biochars in an acidified soil increased the soil pH, CEC, SOM, as well as plant biomass [93]. Biochar application increases soil charge and cation exchange capacity (CEC) by 20–40% compared to control conditions [94]. Non-wood feedstock biochar exhibits higher CEC than wood-derived biochars [95, 96]. Feedstock type, application rate, and pyrolysis temperature are crucial factors influencing soil CEC regulation [97].

3.6 Biochar application on soil EC and pH modulation

Soil acidity is a significant challenge for upland agriculture, affecting approximately 30% of potential arable lands globally [98]. Factors such as aluminum (Al) toxicity, as well as deficiencies in calcium (Ca), magnesium (Mg), and phosphorus (P), act as constraints on crop production in acidic soils [99]. The alkaline properties of specific biochars play a crucial role in enhancing crop productivity in acidic and highly weathered soils [100]. However, the addition of pine sawdust biochar in sandy desert soil has been observed to decrease soil pH [101]. Hence (**Figure 1**),

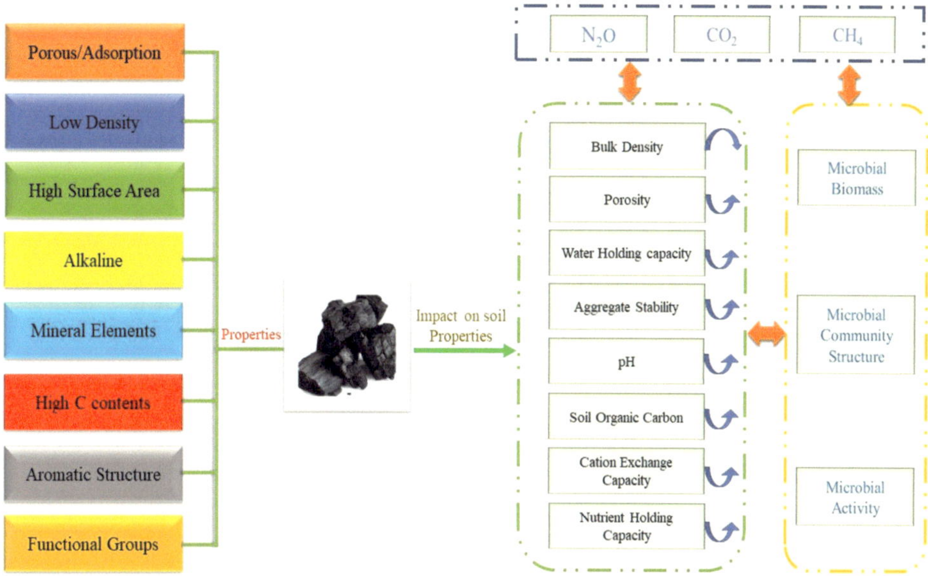

Figure 1.
Conceptual scheme on the effect of biochar amendment on the physicochemical, biological properties, and greenhouse gas emissions (adapted from Li et al. [102, 103]).

it is imperative to exercise caution in selecting the appropriate acidic or alkaline biochar capable of modifying the soil rhizosphere accordingly for optimal plant growth. Electrical conductivity (EC) is closely related to the concentration of salts dissolved in water and is a critical factor influencing upland crop growth (**Table 1**). Biochar derived from agricultural waste and woody feedstock typically exhibits low to moderate EC, while manure-derived biochar tends to have higher EC [116–118]. Biochar application has been shown to improve the growth and yield of crops in saline conditions, attributed to higher concentrations of essential ions in the biochar [118]. Most biochars have higher soluble salt content and, consequently, higher EC compared to agricultural soils [119]. However, excessive salt content is detrimental to plants due to reduced osmotic potential. Therefore, maintaining low soil EC is crucial for optimal nutrient availability and plant growth. It is worth noting that the EC of soil may increase with higher application rates of biochar [102, 103, 120, 121]. Rice husk biochar, however, did not significantly impact soil EC [122]. Biochar pH ranges from 4 to 12 and directly influences soil pH upon application. Biochar interacts with aluminum ions (Al^{3+}) and hydrogen ions (H^+) in soil, leading to reduced ion concentrations [123]. Higher pyrolytic temperatures can result in biochar with alkaline pH [124]. In acidic soils, biochar additions raise soil pH to varying degrees [92, 125, 126]. The pH increases gradually with higher biochar application rates, but the impact on alkali soils is less pronounced [94]. Biochar additions lead to a 25% increase in crop yield in tropical regions, attributed to lower acidity, whereas effects are less significant in temperate zones [127]. Biochar addition also affects soil pH and the activity of essential ions involved in phosphorus (P) complexation and sorption [128]. Research by Lu et al. [129] demonstrated that various types of biochars enhance soil pH and buffering capacity, correlating with improved soil pH buffering capacity.

Experiment	Feedstock	Temperature (° C)	Application rate	Plant	Effect on soil properties	References
Pot	Rice husk	750	0, 2, and 4% w/w	*Oryza sativa*	Improved soil properties observed	[104]
Pot	Woodchip biochar, straw biochar, and vineyard-pruning biochars	525	3% w/w	*Sinapis alba*, *Hordeum vulgare*, and *Trifolium pratense*	Lowered bulk density and enhanced aggregate stability	[63]
Field	Poultry manure	450	5 t ha^{-1}	*Zeya mays*	Increased water retention (3.3–31.3%) after biochar application	[105]
Pot	Coffee ground and coffee husk	530	4, 8, 12, and 16 Mg ha^{-1}	*Zeya mays*	Improved nutrient and water retention, boosting water use efficiency by 50% and enhancing carbon content	[106]
Pot	Sludge and straw biochars	500	2 and 4% w/w	*Triticumaestivum*	Decreased bulk density by 17–18%, tensile strength, and soil surface cracks and increased shear strength.	[107]
Field	Barley straw	400	10 t ha^{-1}	*Brassica rapa*	Decreased the bulk density, enhanced soil porosity and s pH, total N, CEC, and available P	[108]
Field	Cattle manure	600	0, 1.5, 2.5, and 5% w/w	*Spinacia oleracea*	Increased total porosity by 2–12%, and water use efficiency	[109]
Pot	Wood, bamboo, rice straw, and Chinese walnut shell	500	5% w/w	*Phyllostachy pubescens*	Improved the soil EC, and decreased the soil pH.	[110]
Pot	Soft wood biochar made by gasification procedure	500 and 600	4 and 10% w/w	*Lollium perenne* and *Eruca sativa*	Increased water retention decreased in the pH	[111]
Field	Peanut shell	220	5 t ha^{-1}	*Nicotiana tabacum*	Decline the bulk density, increased total N, total porosity, water contents, and nutrient holding capacity	[112]

Experiment	Feedstock	Temperature (°C)	Application rate	Plant	Effect on soil properties	References
Greenhouse	Wheat straw	350–550	0, 25, and 50 t ha^{-1}	*Solanum lycopersicum L.*	Decreased the bulk density, enhanced soil porosity, and improved soil 3-phase composition	[44–46]
Greenhouse	Southern yellow pine	400	5, 10, 15, and 20% w/w	*Vitis rotundifolia*	Decreased the bulk density (40%), increased soil porosity (50%) and soil pH buffering capacity	[113]
Field	Rice husk	—	0 and 2 t ha^{-1}	*Brassica oleracea*	Improved infiltration rate, hydraulic conductivity, maximum water holding capacity, and aggregate stability.	[114]
Pot	Maize residues	450–950	0, 2.1, 4.2, and 8.3 g kg^{-1} w/w	*Phaseolusmultiflorus*	Improved the soil aggregation (188%) and soil water retention (128.9%)	[115]

Table 1.
Effect of biochar addition on soil physicochemical properties and plant growth.

3.7 Improvement in soil organic matter (SOM)

Soil organic matter (SOM) is a vital factor reflecting soil fertility and providing nutrients for soil life [130]. Biochar application increases SOM, with effectiveness dependent on biochar quantity and stability [131, 132]. Biochar contains essential plant nutrients, enriching soil nutrient levels upon application [57]. It also catalyzes the formation of soil organic matter from small organic molecules [82, 133]. Biochar enhances soil fertility and crop yields by boosting SOM content [134]. Elevated SOM content improves crop growth and productivity due to increased soil porosity and nutrient availability [135]. In coastal regions, biochar.

addition increases corn yields while reducing soil salinity [136]. It also enhances root microbe activity and organic matter uptake by plant roots. Chen et al. [133] found that applying biochar at 15 and 30 t ha^{-1} increased SOM content. Additionally, Liu et al. [137] showed that biochar application reduced organic matter mineralization by inhibiting the β-glucosidase enzyme.

3.8 Improvement in soil biological properties

The effects of biochar application on soil microbial biomass are diverse, with some studies indicating positive impacts [138] and others showing no significant changes [139]. Additionally, biochar can enhance soil microbial biomass carbon, nitrogen, C-mineralization, and enzyme activities [140, 141]. Barman et al. [142] observed an increase in different enzyme types with freshly prepared biochar. Moreover, biochar has been found to mitigate toxic contaminants and modify soil enzymes [143]. Furthermore, biochar application has been associated with increased root colonization and spore germination of mycorrhizal fungus, attributing to improvements in soil properties [144, 145]. It has also shown potential to enhance biological nitrogen fixation in various crops [146–148]. Shifts in microbial community structures and functions after biochar supplementation are influenced by physicochemical attributes of biochar, organic matter content, nutrient availability, and water retention capacity [149]. Additionally, dominant bacterial groups like *Alphaproteobacteria, Betaproteobacteria, Proteobacteria, Gammaproteobacteria, and Rubrobacteridae* have been reported to increase after biochar addition [150]. Tang et al. [151] observed that soil microorganisms, including those in the rhizosphere, utilize various carbon sources after biochar amendment. In a 30-month experiment, Somerville et al. [152] found that the addition of organic matter, including biochar, improved soil biological properties, attributing it to increased porosity supporting microbial biomass, particularly mycorrhiza [153]. Kolton et al. [154] noted that biochar has the potential to enhance the diversity and metabolic potential of microbiota in the tomato rhizosphere. Pei et al. [155] reported that biochar amendment increased C mineralization, microbial respiration, and microbial communities. Additionally, nutrient-enriched biochar stimulated microbial activities responsible for pesticide degradation [156, 157], indicating a close interaction between microbial activities and pesticides in soils [158]. Biochar application also enhanced the alpha diversity of various microbial communities, particularly fungal species [159]. While numerous studies have focused on improving soil biological attributes with biochar application (see **Table 2**), there is still much to learn about the dynamics of soil microbial communities and the enhancement of soil biological properties in relation to biochar application.

Experiment	Feedstock	Temperature (° C)	Application rate	Plant	Effect on biological properties	References
Pot trial	Pine needle and *Lantana* biochar	—	2 and 5 t C ha^{-1}	*Triticumaestivum* and *Oryza sativa*	Increment in the soil biological activities	[160]
Field	Red gram, Maize stalk, cotton stalk, and mesquite wood	350–400	2.5 and 5 t ha^{-1}	*Arachishypogaea*	Increment in the population of actinomycetes, bacteria, and fungi	[161]
Field	Poultry litter	300	2.25 and 5 t DM ha^{-1}	Pasture grass	Reduced soil toxicity to *Heterocyprisincongruens*and *Vibrio fischeri*	[162]
Pot	Mixture of maple, oak and birch woody biomass	—	1 t ha^{-1}	*Zea mays*	Increment in the soil potential microbial activities	[163]
Field and white peat	Holm oak	650	3% w/w	*Lactuca sativa* and *Fragariaananassa*	In strawberry grown in white peat, biochar application altered the rhizosphere microbiology	[164]
Field	Local timber harvest	450–550	20 t ha^{-1}	*Cucurbita maxima*	Increased microbial biomass C, enzyme activities, and phosphatase enzyme	[165]
Pot	Orchard pruning biomass	500	65 g kg^{-1}	*Lactuca sativa*	Biochar in combination with compost enhanced the population and diversity of microorganisms	[166]
Field	Wood biochar	450–550	20 t ha^{-1}	—	A shift in the bacterial dominated microbial communities was observed after the biochar addition for 3 months	[167]
Pot	Hardwood lumber scraps	500–550	20% w/w	*Solanum melongena*	Biochar alone and in combination with compost altered the soil microbial communities and their functional diversity	[168]
Growth chamber/Pot trial	Maize	210 and 600	2% w/v	*Cicer arietinum*	Improved the symbiotic relationship with *Mesorhizobiumciceri*and improved the nodule number by 52%	[169]
Field	Rice straw	450	1500 kg ha^{-1} a^{-1}	*Oryza sativa*	Improved the microbial biomass C and microbial biomass N	[170]

Experiment	Feedstock	Temperature (° C)	Application rate	Plant	Effect on biological properties	References
Field	Waste eucalypt wood	550	10% v/v	*Corymbia maculata*	Improved microbially mediated OM decomposition, increased microbial activities	[152]
Pot	*Platanus orientalis* derived	650	0, 0.5, 1, 2, and 4% w/w	*Brassica chinensis*	Increased sucrase, urease, and catalase activities	[171]
Field	Yellow pine	350	11.2 and 22.4 t ha^{-1}	*Paspalum distichum*	Compost amendment delivered more promising results as compared to biochar amendments in terms of microbial biomass	[172]
Pot trial	Cherrywood, wood, and maize	450, 600, and 850	2% w/w	*Lactuca sativa*	Improved soil enzyme activities (N and P cycling) and plant interaction with microbial inoculants	[173]
Field	*Senna siamea*	485	2000 kg ha^{-1}	*Zeya mays*	Improvement in the dehydrogenase activities, fungal, bacterial population and microbial biomass C	[174]

Table 2.
Effect of biochar prepared from different sources on the soil biological properties.

4. Effect of biochar amendment on soil fertility

Soil fertility refers to a soil's ability to provide essential plant nutrients for sustained crop productivity [175]. Modern agricultural practices often lead to soil degradation and nutrient depletion [175]. Biochar addition improves soil properties by increasing organic carbon and crucial mineral nutrients like nitrogen, phosphorus, potassium, calcium, magnesium, and sulfur [176]. However, biochar's nutrient composition depends on raw material source and pyrolysis conditions, which may not always meet diverse soil requirements [177]. Biochar also enhances nutrient retention and utilization in soil. After incorporation, nitrate and ammonium nitrogen retention increases significantly [178]. Additionally, biochar treatment leads to substantial increases in grain biomass and nitrogen and phosphorus utilization efficiency compared to using nitrogen fertilizer alone [45]. Research on biochar production and applications is rapidly expanding [179, 180]. While carbon, nitrogen, and phosphorus cycling influences soil microbial activity, other essential elements like sulfur, potassium, magnesium, and calcium are crucial for soil fertility and quality [181, 182]. Biochars rich in exchangeable base cations support plant growth [183], and a strong correlation exists between oilseed rape yield and soil potassium content [184]. Comprehensive assessment of soil fertility considers interconnected interactions, with integrative indicators like soil pH, electrical conductivity, and cation exchange capacity providing a more holistic understanding [181, 182].

5. Improvement in carbon (C) sequestration

The rise in ambient temperature due to anthropogenic activities has led to a 0.88°C increase in global surface temperature since the nineteenth century [185, 186]. Biochar application offers an eco-friendly solution for soil organic carbon (SOC) sequestration [11, 187–189]. It enhances SOC recalcitrance under warming conditions, promoting carbon sequestration [190]. Biochar also inhibits SOC mineralization by increasing biochemical recalcitrance or microbial activity [191, 192]. Field experiments over three years demonstrate significant improvements in soil carbon sequestration with biochar addition [193]. Various studies confirm biochar's efficacy in enriching the SOC pool and reducing GHG emissions in agricultural soils [194–198]. Feedstock and pyrolysis conditions are critical factors influencing biochar's effectiveness [199]. Zhang et al. [200] found that biochar addition increases recalcitrant carbon in soil, reducing carbon cycling enzyme activity while enhancing phosphorus, nitrogen cycling enzyme, and oxidase activities. They suggest combining biochar with nitrogen fertilization for enhanced carbon sequestration. Rahman et al. [201] report that biochar application from sugarcane bagasse improves SOC retention and other soil quality parameters. Han et al. [38] demonstrate significant increases in carbon sequestration and reductions in net global warming potential with apple wood-derived biochar application, offering promise for mitigating global warming concerns related to apple production.

Author details

Hassan Ali[1], Shahzaib Ali[2], Sadia Baloch[2], Fahmeeda Naheed[1], Emaan Amjad[1], Qudsia Saeed[3], Muhammad Naveed[1] and Adnan Mustafa[3,4,5]*

1 Institute of Soil and Environmental Sciences, University of Agriculture, Faisalabad, Pakistan

2 Department of Agroecosystems, University of South Bohemia, Ceske Budejovice, Czech Republic

3 Faculty of Chemistry, Institute of Chemistry and Technology of Environmental Protection, Brno University of Technology, Brno, Czech Republic

4 Department of Agrochemistry, Soil Science, Microbiology and Plant Nutrition, Faculty of AgriSciences, Mendel University in Brno, Brno, Czech Republic

5 Faculty of Science, Institute for Environmental Studies, Charles University in Prague, Praha, Czech Republic

*Address all correspondence to: adnanmustafa780@gmail.com

References

[1] Bai L, Wang C, Zang S, Wu C, Luo J, Wu Y. Mapping Soil alkalinity and salinity in northern Songnen Plain, China with the HJ-1 hyperspectral imager data and partial least squares regression. Sensors. 2018;**18**:3855. DOI: 10.3390/s18113855

[2] Panwar NL, Pawar A, Salvi BL. Comprehensive review on production and utilization of biochar SN. Applied Sciences. 2019;**1**:168. DOI: 10.1007/s42452-019-0172-6

[3] Hou D, Bolan NS, Tsang DCW, Kirkham MB, O'Connor D. Sustainable soil use and management: An interdisciplinary and systematic approach. Science of the Total Environment. 2020;**729**:138961

[4] Priit T, Ana Catarina B, Simon J, Fr'ed'eric R, Jürgen K, Ellen RG, et al. Biochars in soils: Towards the required level of scientific understanding. Journal of Environmental Engineering and Landscape Management. 2017;**25**:2

[5] Annabi M, Le Bissonnais Y, Le Villio-Poitrenaud M, Houot S. Improvement of soil aggregate stability by reported applications of organic amendments to a cultivated silty loam soil. Agriculture, Ecosystems & Environment. 2011;**144**:382-389

[6] De Meyer A, Poesen J, Isabirye M, Deckers J, Rates D. Soil erosion rate in tropical villages: A case study from Lake Victoria Basin, Uganda. Catena. 2011;**84**:89-98

[7] Drobnik T, Greiner L, Keller A, Grˆet-Regamey A. Soil quality indicators – From soil functions to ecosystem services. Ecological Indicators. 2018;**94**:151-169

[8] Gupta UC, Gupta SC. Quality of animal and human life as affected by selenium management of soils and crops. Communications in Soil Science and Plant Analysis. 2002;**33**(15-18):2537-2555

[9] Khalid S, Shahid M, Murtaza B, Bibi I, Naeem MA, Niazi NK. A critical review of different factors governing the fate of pesticides in soil under biochar application. Science of the Total Environment. 2020;**711**:134645

[10] Agyarko-Mintah E, Cowie A, Van Zwieten L, Singh BP, Smillie R, Harden S, et al. Biochar lowers ammonia emission and improves nitrogen retention in poultry litter composting. Waste Management. 2017;**61**:129-137

[11] Lehmann J, Rillig MC, Thies J, Masiello CA, Hockaday WC, Crowley D. Biochar effects on soil biota–A review. Soil Biology and Biochemistry. 2011;**43**(9):1812-1836

[12] Obia A, Cornelissen G, Martinsen V, Smebye AB, Mulder J. Conservation tillage and biochar improve soil water content and moderate soil temperature in a tropical Acrisol. Soil and Tillage Research. 2020;**197**:104521

[13] Yan Q, Dong F, Li J, Duan Z, Yang F, Li X, et al. Effects of maize straw-derived biochar application on soil temperature, water conditions and growth of winter wheat. European Journal of Soil Science. 2019;**70**:1280-1289

[14] Xiao X, Chen B, Chen Z, Zhu L, Schnoor JL. Insight into multiple and multilevel structures of biochar sand their potential environmental applications: A critical review. Environmental Science & Technology. 2018;**52**(9):5027-5047

[15] Hafeez A, Pan T, Tian J, Cai K. Modified biochars and their effects on soil quality: A review. Environments. 2022;**9**(5):60

[16] Lehmann JA. Handful of carbon. Nature. 2007;**447**:143-144

[17] Nair VD, Nair P, Dari B, Freitas AM, Chatterjee N, Pinheiro FM. Biochar in the agroecosystem–climate-change–sustainability nexus. Frontiers in Plant Science. 2017;**8**:2051

[18] Hairani A, Osaki M, Watanabe T. Effect of biochar application on mineral and microbial properties of soils growing different plant species. Soil Science & Plant Nutrition. 2016;**62**:519-525

[19] Jeffery S, Verheijen FG, Kammann C, Abalos D. Biochar effects on methane emissions from soils: A meta-analysis. Soil Biology and Biochemistry. 2016;**101**:251-258

[20] Xia S, Song Z, Jeyakumar P, Bolan N, Wang H. Characteristics and applications of biochar for remediating Cr(VI)-contaminated soils and wastewater. Environmental Geochemistry and Health. 2020;**42**(6):1543-1567

[21] Zhang H, Voroney R, Price G. Effects of temperature and processing conditions on biochar chemical properties and their influence on soil C and N transformations. Soil Biology and Biochemistry. 2015;**83**:19-28

[22] Blanco-Canqui H. Biochar and soil physical properties. Soil Science Society of America Journal. 2017;**81**(4):687-711

[23] Rawat J, Saxena J, Sanwal P. Biochar: A sustainable approach for improving plant growth and soil properties. In: Biochar—An Imperative Amendment for Soil and the Environment. London, UK: IntechOpen; 2019. p. 10

[24] Shakya A, Agarwal T. Potential of biochar for the remediation of heavy metal contaminated soil. In: Biochar Applications in Agriculture and Environment Management. Berlin/Heidelberg, Germany: Springer; 2020. pp. 77-98

[25] Yu H, Zou W, Chen J, Chen H, Yu Z, Huang J, et al. Biochar amendment improves crop production in problem soils: A review. Journal of Environmental Management. 2019;**232**:8-21

[26] Zheng H, Wang X, Chen L, Wang Z, Xia Y, Zhang Y, et al. Enhanced growth of halophyte plants in biochar-amended coastal soil: Roles of nutrient availability and rhizosphere microbial modulation. Plant, Cell & Environment. 2018;**41**(3):517-532

[27] Zheng H, Wang X, Luo X, Wang Z, Xing B. Biochar-induced negative carbon mineralization priming effects in a coastal wetland soil: Roles of soil aggregation and microbial modulation. Science of the Total Environment. 2018;**610**:951-960

[28] Safadoust A, Feizee P, Mahboubi AA, Gharabaghi B, Mosaddeghi MR, Ahrens B. Least limiting water range as affected by soil texture and cropping system. Agricultural Water Management. 2014;**136**:34-41

[29] Tormena CA, Karlen DL, Logsdon S, Cherubin MR. Corn Stover harvest and tillage impacts on near-surface soil physical quality. Soil and Tillage Research. 2017;**166**:122-130

[30] Alkharabsheh HM, Seleiman MF, Battaglia ML, Shami A, Jalal RS, Alhammad BA, et al. Biochar and its broad impacts in soil quality and fertility, nutrient leaching and crop productivity: A review. Agronomy. 2021;**11**(5):993

[31] Wang Y, Yin R, Liu R. Characterization of biochar from fast pyrolysis and its effect on chemical properties of the tea garden soil. Journal of Analytical and Applied Pyrolysis. 2014;**110**:375-381

[32] Jien SH, Wang CS. Effects of biochar on soil properties and erosion potential in a highly weathered soil. Catena. 2013;**110**:225-233

[33] Jones DL, Rousk J, Edwards-Jones G, DeLuca TH, Murphy DV. Biochar-mediated changes in soil quality and plant growth in a three year field trial. Soil Biology and Biochemistry. 2012;**45**:113-124

[34] Jia Y, Hu Z, Ba Y, Qi W. Application of biochar-coated urea controlled loss of fertilizer nitrogen and increased nitrogen use efficiency. Chemical and Biological Technologies in Agriculture. 2021;**8**(1):1-11

[35] Berisso FE, Schjønning P, Keller T, Lamandé M, Etana A, De Jonge LW, et al. Persistent effects of subsoil compaction on pore size distribution and gas transport in a loamy soil. Soil and Tillage Research. 2012;**122**:42-51

[36] de Lima RP, da Silva AP, Giarola NF, da Silva AR, Rolim MM. Changes in soil compaction indicators in response to agricultural field traffic. Biosystems Engineering. 2017;**162**:1-10

[37] Emmet-Booth JP, Holden NM, Fenton O, Bondi G, Forristal PD. Exploring the sensitivity of visual soil evaluation to traffic-induced soil compaction. Geoderma Regional. 2020;**20**:e00243

[38] Han L, Zhang B, Chen L, Feng Y, Yang Y, Sun K. Impact of biochar amendment on soil aggregation varied with incubation duration and biochar pyrolysis temperature. Biochar. 2021;**3**(3):339-347

[39] Hannet G, Singh K, Fidelis C, Farrar MB, Muqaddas B, Bai SH. Effects of biochar, compost, and biochar-compost on soil total nitrogen and available phosphorus concentrations in a corn field in Papua New Guinea. Environmental Science and Pollution Research. 2021;**28**(21):27411-27419

[40] İlay REMZİ, Kavdır Y, Memici M, Ekinci K. Grain size-induced changes in carbon and nitrogen concentrations and characteristics of tomato harvest residue biochar. International journal of Environmental Science and Technology. 2020;**17**(9):3917-3926

[41] Wang YP, Liu YL, Tian SQ, Yang JJ, Wang L, Ma J. Straw biochar enhanced removal of heavy metal by ferrate. Journal of Hazardous Materials. 2021;**416**:126128

[42] Ding Y, Liu Y, Liu S, Li Z, Tan X, Huang X, et al. Biochar to improve soil fertility. A review. Agronomy for Sustainable Development. 2016;**36**(2):1-18

[43] Lal R. Soil health and carbon management. Food and Energy Security. 2016;**5**:212-222

[44] Zhang C, Li X, Yan H, Ullah I, Zuo Z, Li L, et al. Effects of irrigation quantity and biochar on soil physical properties, growth characteristics, yield and quality of greenhouse tomato. Agricultural Water Management. 2020;**241**:106263

[45] Zhang Q, Song Y, Wu Z, Yan X, Gunina A, Kuzyakov Y, et al. Effects of six-year biochar amendment on soil aggregation, crop growth, and nitrogen and phosphorus use efficiencies in a rice-wheat rotation. Journal of Cleaner Production. 2020;**242**:118435

[46] Zhang Q, Xiao J, Xue J, Zhang L. Quantifying the effects of biochar application on greenhouse gas emissions from agricultural soils: A global meta-analysis. Sustainability. 2020;**12**:3436

[47] IPCC. Climate Change 2014: Synthesis Report. Contribution of Working Groups I, II and III to the Fifth Assessment Report of the Intergovernmental Panel on Climate Change. Geneva, Switzerland: IPCC; 2014. ISBN 9789291691432

[48] Bolan N, Hoang SA, Beiyuan J, Gupta S, Hou D, Karakoti A, et al. Multifunctional applications of biochar beyond carbon storage. International Materials Reviews. 2022;**67**(2):150-200

[49] Kwon G, Bhatnagar A, Wang H, Kwon EE, Song H. A review of recent advancements in utilization of biomass and industrial wastes into engineered biochar. Journal of Hazardous Materials. 2020;**400**:123242

[50] Lehmann J, Cowie A, Masiello CA, Kammann C, Woolf D, Amonette JE, et al. Biochar in climate change mitigation. Nature Geoscience. 2021;**14**(12):883-892

[51] Osman AI, Hefny M, Abdel Maksoud MIA, Elgarahy AM, Rooney DW. Recent advances in carbon capture storage and utilisation technologies: A review. Environmental Chemistry Letters. 2021;**19**(2):797-849

[52] Brassard P, Godbout S, Raghavan V. Soil biochar amendment as a climate change mitigation tool: Key parameters and mechanisms involved. Journal of Environmental Management. 2016;**181**:484-497

[53] Qambrani NA, Rahman MM, Won S, Shim S, Ra C. Biochar properties and eco-friendly applications for climate change mitigation, waste management, and wastewater treatment: A review. Renewable and Sustainable Energy Reviews. 2017;**79**:255-273

[54] Stewart CE, Zheng J, Botte J, Cotrufo MF. Co-generated fast pyrolysis biochar mitigates greenhouse gas emissions and increases carbon sequestration in temperate soils. GCB Bioenergy. 2013;**5**(2):153-163

[55] Windeatt JH, Ross AB, Williams PT, Forster PM, Nahil MA, Singh S. Characteristics of biochars from crop residues: Potential for carbon sequestration and soil amendment. Journal of Environmental Management. 2014;**146**:189-197

[56] Zhang X, Xia YS, Shang YJ, Zhao QQ, Shi J. Effects of biochar (BC) on microbial diversity of cadmium (Cd) contaminated soil. China Environmental Science. 2017;**37**(1):252-262

[57] Zhang Y, Wang J, Feng Y. The effects of biochar addition on soil physicochemical properties: A review. Catena. 2021;**202**:105284

[58] Alghamdi AG. Biochar as a potential soil additive for improving soil physical properties—A review. Arabian Journal of Geosciences. 2018;**11**(24):1-16

[59] Ahmad Bhat S, Kuriqi A, Dar MUD, Bhat O, Sammen SS, Towfiqul Islam ARM, et al. Application of biochar for improving physical, chemical, and hydrological soil properties: A systematic review. Sustainability. 2022;**14**(17):11104

[60] Jain S, Singh A, Khare P, Chanda D, Mishra D, Shanker K, et al. Toxicity assessment of *Bacopamonnieri* L. grown in biochar amended extremely acidic coal mine spoils. Ecological Engineering. 2017;**108**:211-219

[61] Liu Y, Zhu J, Ye C, Zhu P, Ba Q, Pang J, et al. Effects of biochar application on the abundance and community composition of denitrifying bacteria in a reclaimed soil from coal mining subsidence area. Science of the Total Environment. 2018;**625**:1218-1224

[62] Villagra-Mendoza K, Horn R. Effect of biochar addition on hydraulic functions of two textural soils. Geoderma. 2018;**326**:88-95

[63] Burrell LD, Zehetner F, Rampazzo N, Wimmer B, Soja G. Long-term effects of biochar on soil physical properties. Geoderma. 2016;**282**:96-102

[64] Chen C, Zhu H, Lv Q, Tang Q. Impact of biochar on red paddy soil physical and hydraulic properties and rice yield over 3 years. Journal of Soils and Sediments. 2022;**22**(2):607-616

[65] Głąb T, Palmowska J, Zaleski T, Gondek K. Effect of biochar application on soil hydrological properties and physical quality of sandy soil. Geoderma. 2016;**281**:11-20

[66] Qin X, Wang H, Liu C, Li J, Wan Y, Gao Q, et al. Long-term effect of biochar application on yield-scaled greenhouse gas emissions in a rice paddy cropping system: A four-year case study in South China. Science of the Total Environment. 2016;**569**:1390-1401

[67] Verheijen FG, Zhuravel A, Silva FC, Amaro A, Ben-Hur M, Keizer JJ. The influence of biochar particle size and concentration on bulk density and maximum water holding capacity of sandy vs sandy loam soil in a column experiment. Geoderma. 2019;**347**:194-202

[68] Hina K, Bishop P, Arbestain MC, Calvelo-Pereira R, Maciá-Agulló JA, Hindmarsh J, et al. Producing biochars with enhanced surface activity through alkaline pretreatment of feedstocks. Soil Research. 2010;**48**(7):606-617

[69] Suliman W, Harsh JB, Abu-Lail NI, Fortuna AM, Dallmeyer I, Garcia-Pérez M. The role of biochar porosity and surface functionality in augmenting hydrologic properties of a sandy soil. Science of the Total Environment. 2017;**574**:139-147

[70] Bonelli PR, Buonomo EL, Cukierman AL. Pyrolysis of sugarcane bagasse and co-pyrolysis with an Argentinean subbituminous coal. Energy Sources, Part A: Recovery, Utilization, and Environmental Effects. 2007;**29**(8):731-740

[71] Tomczyk A, Sokołowska Z, Boguta P. Biochar physicochemical properties: Pyrolysis temperature and feedstock kind effects. Reviews in Environmental Science and Bio/Technology. 2020;**19**(1):191-215

[72] Wani I, Kushvaha V, Garg A, Kumar R, Naik S, Sharma P. Review on effect of biochar on soil strength: Towards exploring usage of biochar in geo-engineering infrastructure. Biomass Conversion and Biorefinery. 2022:1-32. DOI: 10.1007/s13399-022-02795-5

[73] Yang C, Liu J, Ying H, Lu S. Soil pore structure changes induced by biochar affect microbial diversity and community structure in an Ultisol. Soil and Tillage Research. 2022;**224**:105505

[74] Negassa WC, Guber AK, Kravchenko AN, Marsh TL, Hildebrandt B, Rivers ML. Properties of soil pore space regulate pathways of plant residue decomposition and community structure of associated bacteria. PLoS One. 2015;**10**(4):e0123999

[75] Seyedsadr S, Šípek V, Jačka L, Sněhota M, Beesley L, Pohořelý M, et al.

Biochar considerably increases the easily available water and nutrient content in low-organic soils amended with compost and manure. Chemosphere. 2022;**293**:133586

[76] Hardie M, Clothier B, Bound S, Oliver G, Close D. Does biochar influence soil physical properties and soil water availability? Plant and Soil. 2014;**376**(1):347-361

[77] Cañasveras JC, Barrón V, Del Campillo MC, Torrent J, Gómez JA. Estimation of aggregate stability indices in Mediterranean soils by diffuse reflectance spectroscopy. Geoderma. 2010;**158**(1-2):78-84

[78] Yan FL, Shi ZH, Li ZX, Cai CF. Estimating interrill soil erosion from aggregate stability of Ultisols in subtropical China. Soil and Tillage Research. 2008;**100**(1-2):34-41

[79] Glaser B, Lehmann J, Zech W. Ameliorating physical and chemical properties of highly weathered soils in the tropics with charcoal–A review. Biology and Fertility of Soils. 2002;**35**(4):219-230

[80] Sun F, Lu S. Biochars improve aggregate stability, water retention, and pore-space properties of clayey soil. Journal of Plant Nutrition and Soil Science. 2014;**177**(1):26-33

[81] Wang D, Fonte SJ, Parikh SJ, Six J, Scow KM. Biochar additions can enhance soil structure and the physical stabilization of C in aggregates. Geoderma. 2017;**303**:110-117

[82] Liang B, Lehmann J, Sohi SP, Thies JE, O'Neill B, Trujillo L, et al. Black carbon affects the cycling of non-black carbon in soil. Organic Geochemistry. 2010;**41**(2):206-213

[83] Yuan Z, Song Y, Li D, Huang B, Chen Y, Ge X, et al. Effects of biochar application on the loss characteristics of Cd from acidic soil under simulated rainfall conditions. Environmental Science and Pollution Research. 2022;**30**(2):1-12

[84] Wang H, She D, Fei Y, Tang S. Synergic effects of biochar and polyacrylamide amendments on the mechanical properties of silt loam soil under coastal reclamation in China. Catena. 2019;**182**:104152

[85] Zong Y, Xiao Q, Lu S. Acidity, water retention, and mechanical physical quality of a strongly acidic Ultisol amended with biochars derived from different feedstocks. Journal of Soils and Sediments. 2016;**16**(1): 177-190

[86] Mukome FN, Zhang X, Silva LC, Six J, Parikh SJ. Use of chemical and physical characteristics to investigate trends in biochar feedstocks. Journal of Agricultural and Food Chemistry. 2013;**61**(9):2196-2204

[87] Safari N, Aliasgharzad N, Asghari S. The effects of polyacrylamide on the parameters of physical quality in a clay loam soil selected from semiarid region. Journal of Soil Environment. 2015;**1**(1):10-17

[88] Choudhary MP, Charan HD, Acharya B. Experimental investigation on biochar production and application: Solution to air pollution due to stubble burning as well as amending soil consistency. Journal of The Institution of Engineers (India): Series A. 2021;**102**(3):705-712

[89] Liang B, Lehmann J, Solomon D, Kinyangi J, Grossman J, O'Neill BJOJFJJEG, et al. Black carbon increases cation exchange capacity in soils. Soil Science Society of America Journal. 2006;**70**(5):1719-1730

[90] Hossain MK, Strezov V, Chan KY, Ziolkowski A, Nelson PF. Influence of pyrolysis temperature on production and nutrient properties of wastewater sludge biochar. Journal of Environmental Management. 2011;**92**(1):223-228

[91] Melo LC, Coscione AR, Abreu CA, Puga AP, Camargo OA. Influence of pyrolysis temperature on cadmium and zinc sorption capacity of sugar cane straw–derived biochar. BioResources. 2013;**8**(4):4992-5004

[92] Novak JM, Lima I, Xing B, Gaskin JW, Steiner C, Das KC, et al. Characterization of designer biochar produced at different temperatures and their effects on a loamy sand. Annals of Environmental Science. 2009;**3**(2):195-206

[93] Wang ZY, Chen L, Sun FL, Luo XX, Wang HF, Liu GC, et al. Effects of adding biochar on the properties and nitrogen bioavailability of an acidic soil. European Journal of Soil Science. 2017;**68**(4):559-572

[94] Chintala R, Schumacher TE, McDonald LM, Clay DE, Malo DD, Papiernik SK, et al. Phosphorus sorption and availability from biochars and soil/B iochar mixtures. CLEAN–Soil, Air, Water. 2014;**42**(5):626-634

[95] Carrier M, Hardie AG, Uras Ü, Görgens J, Knoetze JH. Production of char from vacuum pyrolysis of South-African sugar cane bagasse and its characterization as activated carbon and biochar. Journal of Analytical and Applied Pyrolysis. 2012;**96**:24-32

[96] Murtaza G, Ahmed Z, Usman M. Feedstock type, pyrolysis temperature and acid modification effects on physiochemical attributes of biochar and soil quality. Arabian Journal of Geosciences. 2022;**15**(3):1-30

[97] Domingues RR, Sánchez-Monedero MA, Spokas KA, Melo LC, Trugilho PF, Valenciano MN, et al. Enhancing cation exchange capacity of weathered soils using biochar: Feedstock, pyrolysis conditions and addition rate. Agronomy. 2020;**10**(6):824

[98] Fageria NK, Baligar VC. Ameliorating soil acidity of tropical Oxisols by liming for sustainable crop production. Advances in Agronomy. 2008;**99**:345-399

[99] Zheng SJ. Crop production on acidic soils: Overcoming aluminium toxicity and phosphorus deficiency. Annals of Botany. 2010;**106**(1):183-184

[100] Lehmann J, Rondon M. Bio-char soil management on highly weathered soils in the humid tropics. Biological Approaches to Sustainable Soil Systems. 2006;**113**(517):e530

[101] Laghari M, Mirjat MS, Hu Z, Fazal S, Xiao B, Hu M, et al. Effects of biochar application rate on sandy desert soil properties and sorghum growth. Catena. 2015;**135**:313-320

[102] Li C, Xiong Y, Qu Z, Xu X, Huang Q, Huang G. Impact of biochar addition on soil properties and water-fertilizer productivity of tomato in semi-arid region of Inner Mongolia, China. Geoderma. 2018;**331**:100-108

[103] Li Y, Hu S, Chen J, Müller K, Li Y, Fu W, et al. Effects of biochar application in forest ecosystems on soil properties and greenhouse gas emissions: A review. Journal of Soils and Sediments. 2018;**18**(2):546-563

[104] Pratiwi EPA, Shinogi Y. Rice husk biochar application to paddy soil and its effects on soil physical properties, plant growth, and methane emission. Paddy and Water Environment. 2016;**14**(4):521-532

[105] Are KS, Adelana AO, Fademi IO, Aina OA. Improving physical properties of degraded soil: Potential of poultry manure and biochar. Agriculture and Natural Resources. 2017;**51**(6):454-462

[106] de Sousa Lima JR, de Moraes Silva W, de Medeiros EV, Duda GP, Corrêa MM, Martins Filho AP, et al. Effect of biochar on physicochemical properties of a sandy soil and maize growth in a greenhouse experiment. Geoderma. 2018;**319**:14-23

[107] Malik Z, Yutong Z, ShengGao L, Abassi GH, Ali S, Kamran M, et al. Effect of biochar and quicklime on growth of wheat and physicochemical properties of Ultisols. Arabian Journal of Geosciences. 2018;**11**(17):1-12

[108] Kang SW, Kim SH, Park JH, Seo DC, Ok YS, Cho JS. Effect of biochar derived from barley straw on soil physicochemical properties, crop growth, and nitrous oxide emission in an upland field in South Korea. Environmental Science and Pollution Research. 2018;**25**(26):25813-25821

[109] Gavili E, Moosavi AA, MoradiChoghamarani F. Cattle manure biochar potential for ameliorating soil physical characteristics and spinach response under drought. Archives of Agronomy and Soil Science. 2018;**64**(12):1714-1727

[110] Wang Y, Zhong B, Shafi M, Ma J, Guo J, Wu J, et al. Effects of biochar on growth, and heavy metals accumulation of moso bamboo (*Phyllostachy pubescens*), soil physical properties, and heavy metals solubility in soil. Chemosphere. 2019;**219**:510-516

[111] Teodoro M, Trakal L, Gallagher BN, Šimek P, Soudek P, Pohořelý M, et al. Application of co-composted biochar significantly improved plant-growth relevant physical/chemical properties of a metal contaminated soil. Chemosphere. 2020;**242**:125255

[112] Wang H, Yan S, Ren T, Yuan Y, Kuang G, Wang B, et al. Novel environmental factors affecting microbial responses and physicochemical properties by sequentially applied biochar in black soil. Environmental Science and Pollution Research. 2020;**27**(30):37432-37443

[113] Chang Y, Rossi L, Zotarelli L, Gao B, Shahid MA, Sarkhosh A. Biochar improves soil physical characteristics and strengthens root architecture in muscadine grape (*Vitis rotundifolia* L.). Chemical and Biological Technologies in Agriculture. 2021;**8**(1):1-11

[114] Sharma P, Abrol V, Sharma V, Chaddha S, Rao CS, Ganie AQ, et al. Effectiveness of biochar and compost on improving soil hydro-physical properties, crop yield and monetary returns in inceptisol subtropics. Saudi Journal of Biological Sciences. 2021;**28**(12):7539-7549

[115] Kang MW, Yibeltal M, Kim YH, Oh SJ, Lee JC, Kwon EE, et al. Enhancement of soil physical properties and soil water retention with biochar-based soil amendments. Science of the Total Environment. 2022;**836**:155746

[116] Cely P, Gascó G, Paz-Ferreiro J, Méndez A. Agronomic properties of biochars from different manure wastes. Journal of Analytical and Applied Pyrolysis. 2015;**111**:173-182

[117] Masto RE, Ansari MA, George J, Selvi VA, Ram LC. Co-application of biochar and lignite fly ash on soil nutrients and biological parameters at different crop growth stages of *Zea mays*. Ecological Engineering. 2013;**58**:314-322

[118] Usman ARA, Al-Wabel MI, Abdulaziz AH, Mahmoud WA,

El-Naggar AH, Ahmad M, et al. Conocarpus biochar induces changes in soil nutrient availability and tomato growth under saline irrigation. Pedosphere. 2016;**26**(1):27-38

[119] Igalavithana AD, Mandal S, Niazi NK, Vithanage M, Parikh SJ, Mukome FN, et al. Advances and future directions of biochar characterization methods and applications. Critical Reviews in Environmental Science and Technology. 2017;**47**(23):2275-2330

[120] El-Naggar A, Lee SS, Awad YM, Yang X, Ryu C, Rizwan M, et al. Influence of soil properties and feedstocks on biochar potential for carbon mineralization and improvement of infertile soils. Geoderma. 2018;**332**:100-108

[121] Prapagdee S, Tawinteung N. Effects of biochar on enhanced nutrient use efficiency of green bean, *Vigna radiata* L. Environmental Science and Pollution Research. 2017;**24**:9460-9467

[122] Jatav HS, Singh SK, Singh Y, Kumar O. Biochar and sewage sludge application increases yield and micronutrient uptake in rice (*Oryza sativa* L.). Communications in Soil Science and Plant Analysis. 2018;**49**(13):1617-1628

[123] Van Zwieten L, Kimber S, Morris S, Chan KY, Downie A, Rust J, et al. Effects of biochar from slow pyrolysis of papermill waste on agronomic performance and soil fertility. Plant and Soil. 2010;**327**(1):235-246

[124] Yuan JH, Xu RK, Zhang H. The forms of alkalis in the biochar produced from crop residues at different temperatures. Bioresource Technology. 2011;**102**(3):3488-3497

[125] Hossain MK, Strezov V, Chan KY, Nelson PF. Agronomic properties of wastewater sludge biochar and bioavailability of metals in production of cherry tomato (*Lycopersicon esculentum*). Chemosphere. 2010;**78**(9):1167-1171

[126] Nielsen S, Joseph S, Ye J, Chia C, Munroe P, van Zwieten L, et al. Crop-season and residual effects of sequentially applied mineral enhanced biochar and N fertilizer on crop yield, soil chemistry and microbial communities. Agriculture, Ecosystems & Environment. 2018;**255**:52-61

[127] Jeffery S, Abalos D, Prodana M, Bastos AC, Van Groenigen JW, Hungate BA, et al. Biochar boosts tropical but not temperate crop yields. Environmental Research Letters. 2017;**12**(5):053001

[128] Ghodszad L, Reyhanitabar A, Oustan S, Alidokht L. Phosphorus sorption and desorption characteristics of soils as affected by biochar. Soil and Tillage Research. 2022;**216**:105251

[129] Lu HL, Li KW, Nkoh JN, He X, Hong ZN, Xu RK. Effects of the increases in soil pH and pH buffering capacity induced by crop residue biochars on available Cd contents in acidic paddy soils. Chemosphere. 2022;**301**:134674

[130] Cookson WR, Abaye DA, Marschner P, Murphy DV, Stockdale EA, Goulding KW. The contribution of soil organic matter fractions to carbon and nitrogen mineralization and microbial community size and structure. Soil Biology and Biochemistry. 2005;**37**(9):1726-1737

[131] Glaser B, Haumaier L, Guggenberger G, Zech W. The 'TerraPreta' phenomenon: A model for sustainable agriculture in the humid tropics. Naturwissenschaften. 2001;**88**(1):37-41

[132] Zygourakis K. Biochar soil amendments for increased crop yields: How to design a "designer" biochar. AICHE Journal. 2017;**63**(12):5425-5437

[133] Chen H, Yang X, Wang H, Sarkar B, Shaheen SM, Gielen G, et al. Animal carcass-and wood-derived biochars improved nutrient bioavailability, enzyme activity, and plant growth in metal-phthalic acid ester co-contaminated soils: A trial for reclamation and improvement of degraded soils. Journal of Environmental Management. 2020;**261**:110246

[134] Shaaban M, Van Zwieten L, Bashir S, Younas A, Núñez-Delgado A, Chhajro MA, et al. A concise review of biochar application to agricultural soils to improve soil conditions and fight pollution. Journal of Environmental Management. 2018;**228**:429-440

[135] Akhtar SS, Li G, Andersen MN, Liu F. Biochar enhances yield and quality of tomato under reduced irrigation. Agricultural Water Management. 2014;**138**:37-44

[136] Huang M, Zhang Z, Zhu C, Zhai Y, Lu P. Effect of biochar on sweet corn and soil salinity under conjunctive irrigation with brackish water in coastal saline soil. Scientia Horticulturae. 2019;**250**:405-413

[137] Liu M, Zhu J, Yang X, Fu Q, Hu H, Huang Q. Biochar produced from the straw of common crops simultaneously stabilizes soil organic matter and heavy metals. Science of the Total Environment. 2022;**828**:154494

[138] Sheng Y, Zhu L. Biochar alters microbial community and carbon sequestration potential across different soil pH. Science of the Total Environment. 2018;**622**:1391-1399

[139] Ma H, Egamberdieva D, Wirth S, Li Q, Omari RA, Hou M, et al. Effect of biochar and irrigation on the interrelationships among soybean growth, root nodulation, plant P uptake, and soil nutrients in a sandy field. Sustainability. 2019;**11**(23):6542

[140] Das SK, Das SK. Acid sulphate soil: Management strategy for soil health and productivity. Popular Kheti. 2015;**3**(2):2-7

[141] Das SK, Ghosh GK, Avasthe R. Application of biochar in agriculture and environment, and its safety issues. Biomass Conversion and Biorefinery. 2020;**13**:1-11

[142] Barman H, Das SK, Roy A. Future of nano science in technology for prosperity: A policy paper. Nanoscience & Technology. 2018;**5**(1):1-5

[143] Das SK, Mukherjee I, Roy A. Flubendiamide as new generation insecticide in plant toxicology: A policy paper. Advance in Clinical Toxicology. 2017;**2**:100-122

[144] Das SK, Avasthe RK, Singh M, Yadav A. Soil health improvement using biochar application in Sikkim: A success story. Innovative Farming. 2018;**3**(1):48-50

[145] Singh NS, Mukherjee I, Das SK, Varghese E. Leaching of clothianidin in two different Indian soils: Effect of organic amendment. Bulletin of Environmental Contamination and Toxicology. 2018;**100**(4):553-559

[146] Barman H, Das SK, Roy A. Zinc in soil environment for plant health and management strategy. Universal Journal of Agricultural Research. 2018;**6**(5):149-154

[147] Das SK, Avasthe RK. Livelihood improvement of rural tribal farmers through soil health management,

input support system and training-a success story. Innovative Farming. 2017;**2**(3):171-173

[148] Roy A, Singh NU, Tripathi AK, Yumnam A, Sinha PK, Kumar B, et al. Dynamics of pulse production in north-east region of India-a state-wise analysis. Economic Affairs. 2017;**62**(4):655-662

[149] Ma H, Egamberdieva D, Wirth S, Bellingrath-Kimura SD. Effect of biochar and irrigation on soybean-rhizobium symbiotic performance and soil enzymatic activity in field rhizosphere. Agronomy. 2019;**9**(10):626

[150] Abujabhah IS, Bound SA, Doyle R, Bowman JP. Effects of biochar and compost amendments on soil physicochemical properties and the total community within a temperate agricultural soil. Applied Soil Ecology. 2016;**98**:243-253

[151] Tang J, Zhang S, Zhang X, Chen J, He X, Zhang Q. Effects of pyrolysis temperature on soil-plant-microbe responses to *Solidago canadensis* L.-derived biochar in coastal saline-alkali soil. Science of the Total Environment. 2020;**731**:138938

[152] Somerville PD, Farrell C, May PB, Livesley SJ. Biochar and compost equally improve urban soil physical and biological properties and tree growth, with no added benefit in combination. Science of the Total Environment. 2020;**706**:135736

[153] Warnock DD, Lehmann J, Kuyper TW, Rillig MC. Mycorrhizal responses to biochar in soil–concepts and mechanisms. Plant and Soil. 2007;**300**(1):9-20

[154] Kolton M, Graber ER, Tsehansky L, Elad Y, Cytryn E. Biochar-stimulated plant performance is strongly linked to microbial diversity and metabolic potential in the rhizosphere. New Phytologist. 2017;**213**(3):1393-1404

[155] Pei J, Zhuang S, Cui J, Li J, Li B, Wu J, et al. Biochar decreased the temperature sensitivity of soil carbon decomposition in a paddy field. Agriculture, Ecosystems & Environment. 2017;**249**:156-164

[156] Egamberdieva D, Jabbarov Z, Arora NK, Wirth S, Bellingrath-Kimura SD. Biochar mitigates effects of pesticides on soil biological activities. Environmental Sustainability. 2021;**4**(2):335-342

[157] Qiu Y, Pang H, Zhou Z, Zhang P, Feng Y, Sheng GD. Competitive biodegradation of dichlobenil and atrazine coexisting in soil amended with a char and citrate. Environmental Pollution. 2009;**157**(11):2964-2969

[158] Irfan M, Hussain Q, Khan KS, Akmal M, Ijaz SS, Hayat R, et al. Response of soil microbial biomass and enzymatic activity to biochar amendment in the organic carbon deficient arid soil: A 2-year field study. Arabian Journal of Geosciences. 2019;**12**(3):1-9

[159] Meng L, Sun T, Li M, Saleem M, Zhang Q, Wang C. Soil-applied biochar increases microbial diversity and wheat plant performance under herbicide fomesafen stress. Ecotoxicology and Environmental Safety. 2019;**171**:75-83

[160] Bhattacharjya S, Chandra R, Pareek N, Raverkar KP. Biochar and crop residue application to soil: Effect on soil biochemical properties, nutrient availability and yield of rice (*Oryza sativa* L.) and wheat (*Triticumaestivum* L.). Archives of Agronomy and Soil Science. 2016;**62**(8):1095-1108

[161] Pandian K, Subramaniayan P, Gnasekaran P, Chitraputhirapillai S.

Effect of biochar amendment on soil physical, chemical and biological properties and groundnut yield in rainfedAlfisol of semi-arid tropics. Archives of Agronomy and Soil Science. 2016;**62**(9):1293-1310

[162] Mierzwa-Hersztek M, Gondek K, Baran A. Effect of poultry litter biochar on soil enzymatic activity, ecotoxicity and plant growth. Applied Soil Ecology. 2016;**105**:144-150

[163] Dil M, Oelbermann M, Xue W. An evaluation of biochar pre-conditioned with urea ammonium nitrate on maize (*Zea mays* L.) production and soil biochemical characteristics. Canadian Journal of Soil Science. 2014;**94**(4):551-562

[164] Caroline A, Debode J, Vandecasteele B, D'Hose T, Cremelie P, Haegeman A, et al. Biological, physicochemical and plant health responses in lettuce and strawberry in soil or peat amended with biochar. Applied Soil Ecology. 2016;**107**:1-12

[165] Gao S, Hoffman-Krull K, DeLuca TH. Soil biochemical properties and crop productivity following application of locally produced biochar at organic farms on Waldron Island, WA. Biogeochemistry. 2017;**136**(1):31-46

[166] Trupiano D, Cocozza C, Baronti S, Amendola C, Vaccari FP, Lustrato G, et al. The effects of biochar and its combination with compost on lettuce (*Lactuca sativa* L.) growth, soil properties, and soil microbial activity and abundance. International Journal of Agronomy. 2017;**2017**:1-13

[167] Gao S, DeLuca TH. Wood biochar impacts soil phosphorus dynamics and microbial communities in organically-managed croplands. Soil Biology and Biochemistry. 2018;**126**:144-150

[168] Edenborn SL, Johnson LM, Edenborn HM, Albarran-Jack MR, Demetrion LD. Amendment of a hardwood biochar with compost tea: Effects on plant growth, insect damage and the functional diversity of soil microbial communities. Biological Agriculture & Horticulture. 2018;**34**(2):88-106

[169] Egamberdieva D, Li L, Ma H, Wirth S, Bellingrath-Kimura SD. Soil amendment with different maize biochars improves chickpea growth under different moisture levels by improving symbiotic performance with *Mesorhizobiumciceri* and soil biochemical properties to varying degrees. Frontiers in Microbiology. 2019;**10**:2423

[170] Zheng Y, Han X, Li Y, Yang J, Li N, An N. Effects of biochar and straw application on the physicochemical and biological properties of paddy soils in Northeast China. Scientific Reports. 2019;**9**(1):1-11

[171] Chen S, Qi G, Ma G, Zhao X. Biochar amendment controlled bacterial wilt through changing soil chemical properties and microbial community. Microbiological Research. 2020;**231**:126373

[172] Hale L, Curtis D, Azeem M, Montgomery J, Crowley DE, McGiffen ME Jr. Influence of compost and biochar on soil biological properties under turfgrass supplied deficit irrigation. Applied Soil Ecology. 2021;**168**:104134

[173] Ma H, Shurigin V, Jabborova D, dela Cruz JA, dela Cruz TE, Wirth S, et al. The integrated effect of microbial inoculants and biochar types on soil biological properties, and plant growth of lettuce (*Lactuca sativa* L.). Plants. 2022;**11**(3):423

[174] Phares CA, Amoakwah E, Danquah A, Akaba S, Frimpong KA,

Mensah TA. Improved soil physicochemical, biological properties and net income following the application of inorganic NPK fertilizer and biochar for maize production. Acta Ecologica Sinica. 2022;**42**(4):289-295

[175] Suzuki K, Matsunaga R, Hayashi K, Matsumoto N, Tabo R, Tobita S, et al. Effects of traditional soil management practices on the nutrient status in Sahelian sandy soils of Niger, West Africa. Geoderma. 2014;**223**:1-8

[176] Hossain MZ, Bahar B, Sarkar B, Donne S, Ok YS, Bolan N. Biochar and its importance on nutrient dynamics in soil and plant. Biochar. 2020;**2**:379-420

[177] Guo XX, Liu HT, Zhang J. The role of biochar in organic waste composting and soil improvement: A review. Waste Management. 2020;**102**:884-899

[178] Gao S, Hoffman-Krull K, Bidwell AL, DeLuca TH. Locally produced wood biochar increases nutrient retention and availability in agricultural soils of the San Juan Islands, USA. Agriculture, Ecosystems & Environment. 2016;**233**:43-54

[179] Abhishek K, Shrivastava A, Vimal V, Gupta AK, Bhujbal SK, Biswas JK, et al. Biochar application for greenhouse gas mitigation, contaminants immobilization and soil fertility enhancement: A state-of-the-art review. Science of the Total Environment. 2022;**853**:158562

[180] Yaashikaa PR, Kumar PS, Varjani S, Saravanan A. A critical review on the biochar production techniques, characterization, stability and applications for circular bioeconomy. Biotechnology Reports. 2020;**28**:e00570

[181] Amoah-Antwi C, Kwiatkowska-Malina J, Thornton SF, Fenton O, Malina G, Szara E. Restoration of soil quality using biochar and brown coal waste: A review. Science of the Total Environment. 2020;**722**:137852

[182] Yang M, Mouazen A, Zhao X, Guo X. Assessment of a soil fertility index using visible and near-infrared spectroscopy in the rice paddy region of southern China. European Journal of Soil Science. 2020;**71**(4):615-626

[183] Gondek K, Mierzwa-Hersztek M, Kopeć M, Sikora J, Lošák T, Grzybowski P. Sewage sludge biochar effects on phosphorus mobility in soil and accumulation in plant. Ecological Chemistry and Engineering S. 2019;**26**(2):367-381

[184] Zhu D, Zhang J, Wang Z, Muhammad RK, Lu J, Li X. Soil available potassium affected by rice straw incorporation and potassium fertilizer application under a rice–oilseed rape rotation system. Soil Use and Management. 2019;**35**(3):503-510

[185] Change OC. Intergovernmental panel on climate change. World Meteorological Organization. 2007;**52**:1-43

[186] IPCC. 2007 Climate Change. Climate Change Impacts, Adaptation and Vulnerability; Working Group II. Geneva, Switzerland: IPCC; 2007

[187] Han L, Sun K, Yang Y, Xia X, Li F, Yang Z, et al. Biochar's stability and effect on the content, composition and turnover of soil organic carbon. Geoderma. 2020;**364**:114184

[188] Ippolito JA, Cui L, Kammann C, Wrage-Mönnig N, Estavillo JM, Fuertes-Mendizabal T, et al. Feedstock choice, pyrolysis temperature and type influence biochar characteristics: A comprehensive meta-data analysis review. Biochar. 2020;**2**(4):421-438

[189] Li S, Wang S, Fan M, Wu Y, Shangguan Z. Interactions between biochar and nitrogen impact soil carbon mineralization and the microbial community. Soil and Tillage Research. 2020;**196**:104437

[190] Fang Y, Singh B, Singh BP. Effect of temperature on biochar priming effects and its stability in soils. Soil Biology and Biochemistry. 2015;**80**:136-145

[191] Chen G, Fang Y, Van Zwieten L, Xuan Y, Tavakkoli E, Wang X, et al. Priming, stabilization and temperature sensitivity of native SOC is controlled by microbial responses and physicochemical properties of biochar. Soil Biology and Biochemistry. 2021;**154**:108139

[192] Chen J, Chen D, Xu Q, Fuhrmann JJ, Li L, Pan G, et al. Organic carbon quality, composition of main microbial groups, enzyme activities, and temperature sensitivity of soil respiration of an acid paddy soil treated with biochar. Biology and Fertility of Soils. 2019;**55**(2):185-197

[193] Chen J, Sun X, Zheng J, Zhang X, Liu X, Bian R, et al. Biochar amendment changes temperature sensitivity of soil respiration and composition of microbial communities 3 years after incorporation in an organic carbon-poor dry cropland soil. Biology and Fertility of Soils. 2018;**54**(2):175-188

[194] Dissanayake PD, You S, Igalavithana AD, Xia Y, Bhatnagar A, Gupta S, et al. Biochar-based adsorbents for carbon dioxide capture: A critical review. Renewable and Sustainable Energy Reviews. 2020;**119**:109582

[195] El-Naggar A, El-Naggar AH, Shaheen SM, Sarkar B, Chang SX, Tsang DC, et al. Biochar composition-dependent impacts on soil nutrient release, carbon mineralization, and potential environmental risk: A review. Journal of Environmental Management. 2019;**241**:458-467

[196] El-Naggar A, Lee MH, Hur J, Lee YH, Igalavithana AD, Shaheen SM, et al. Biochar-induced metal immobilization and soil biogeochemical process: An integrated mechanistic approach. Science of the Total Environment. 2020;**698**:134112

[197] Jing F, Sun Y, Liu Y, Wan Z, Chen J, Tsang DC. Interactions between biochar and clay minerals in changing biochar carbon stability. Science of the Total Environment. 2022;**809**:151124

[198] Saffari N, Hajabbasi MA, Shirani H, Mosaddeghi MR, Mamedov AI. Biochar type and pyrolysis temperature effects on soil quality indicators and structural stability. Journal of Environmental Management. 2020;**261**:110190

[199] Mandal S, Sarkar B, Bolan N, Novak J, Ok YS, Van Zwieten L, et al. Designing advanced biochar products for maximizing greenhouse gas mitigation potential. Critical Reviews in Environmental Science and Technology. 2016;**46**(17):1367-1401

[200] Zhang X, Zhang Q, Zhan L, Xu X, Bi R, Xiong Z. Biochar addition stabilized soil carbon sequestration by reducing temperature sensitivity of mineralization and altering the microbial community in a greenhouse vegetable field. Journal of Environmental Management. 2022;**313**:114972

[201] Rahman MA, Jahiruddin M, Kader MA, Islam MR, Solaiman ZM. Sugarcane bagasse biochar increases soil carbon sequestration and yields of maize and groundnut in charland ecosystem. Archives of Agronomy and Soil Science. 2022;**68**(10):1338-1351

Chapter 2

Perspective Chapter: Sample Preparation Techniques for Electrochemical Analysis of Pesticides and Heavy Metals in Environmental and Food Samples

Ranil Clément Tonleu Temgoua,
Sherman Lesly Zambou Jiokeng, Kevin Yemele Tajeu,
Gullit Deffo, Lionnel Averie Vomo, Fabiola Tonleu Dontsi,
Evangeline Njanja and Ignas Kenfack Tonlé

Abstract

The development of an analytical methodology commonly includes sampling and sample pretreatment-preparation. The sample preparation step should provide the analytes (pesticides, heavy metals, drugs, dyes...etc.) in an adequate medium (typically aqueous or non-aqueous solution) to be detected and/or quantified. It is, therefore, necessary to ensure that the sample is homogeneous and free of interferents, as long as the preparation step is the most significant source of error in the analytical method development and is the most time-consuming step especially when solid samples are analyzed. Given its importance, this preparation step has a fundamental importance in the overall analytical method development, mainly when electroanalytical methods are applied. In this chapter, the steps involved in preparing samples for electrochemical analysis will be described in detail. Specifically, we will be focusing on the sample preparation techniques for the electrochemical analysis of pesticides and heavy metals, in environmental and food samples. For non-electrochemical readers, a brief introduction to the most commonly used electroanalytical methods will be described.

Keywords: electrochemistry, pesticides, heavy metals, samples preparation, environment, food

1. Introduction

Nowadays, several methods for the analysis of compounds such as chromatography, mass spectrometry, fluorescence, ultra-violet (UV–Vis) spectroscopy, electrophoresis, electrochemistry, etc. have been elaborated and are still used by researchers. The last mentioned has become the most popular since it has been established to be the discipline at the interface between the branches of chemistry and many other sciences [1]. In a simple word, the relationship between electricity (movement of electrons) and chemistry (transformation) is known as electrochemistry, which is the science of the application of electricity to chemistry [2]. As a science, it is combined with other discipline, showing its high potential in biology (bio-electrochemistry), catalyst (electro-catalyst), kinetic (electro-kinetic), photolysis (photo-electrochemistry), etc. As a technique of analysis, it is coupled to many other classical methods of analysis to form new hybrid methods more performant such UV–Vis/electrochemistry, electrochemistry-liquid chromatography-mass spectrometry (EC-LC–MS) largely used recently [3].

Electrochemistry studies the links between chemistry and electricity. One of the electrochemical techniques is voltammetry, which is very popular for electroanalysis. By applying a potential (or a current), it is possible to measure the resulting current (or potential) [4]. The identify electroanalytical methods generally used from the most sensitive to the less one are anodic stripping voltammetry (ASV) and cathodic stripping voltammetry (CSV), linear sweeping voltammetry (LSV), normal pulse voltammetry (NPV), square-wave voltammetry (SWV), differential pulse voltammetry (DPV), chronoamperometry/chronocoulometry (CA/CC), cyclic voltammetry (CV), and polarography [5]. All of these techniques used a common electroanalytical setup constituted of a cell, a potentiostat, and a computer. The cell contains the supporting electrolyte (SE) with the sample to be investigated, and three conventional electrodes (working electrode WE, reference electrode RE, and auxiliary electrode AE) connected to a potentiostat used to allow the passage of the current between the WE and AE, and to establish the potential between WE and RE. This happens during the transformation/redox mechanism of the analyte, and the responses collected in the potentiostat are transferred to the computer for interpretation [4, 6].

Due to the facility, rapidity, low cost, and performance of electrochemistry to provide information on the electron transfer of the system, several types of analytes can be investigated using electroanalytical method. Pesticides, heavy metals, dyes, food additives, endocrine disruptors, pharmaceutical compounds, etc. are the different family of samples generally analyzed. The sample preparation of the first two listed (pesticides and heavy metals) will be investigated in this work.

Pesticides are substances intended to repel or combat any undesirable plant or animal species, during the production, storage, transport, distribution, or processing of foodstuffs intended for human or animal consumption. They are classified into three main classes which are insecticides, herbicides, and fungicides [7]. The aim of using a pesticide in the agricultural production process is to fight against undesirable plants and to repel fungi, rodents, and insects. Unfortunately, the vast majority end up in the soil, water, and atmosphere, and through the food chain are transferred to humans, affecting their health [8].

Heavy metals are chemical elements characterized by a high atomic mass, with a tendency to be highly toxic to living beings [9]. Heavy metals are pollutants that represent a growing environmental problem that can affect various components of the environment [10]. It has a wide range of applications, including the manufacture of electrical cables and heating pipes, cladding materials for buildings, catalysis, pulp production, batteries, pigments, weights, radiation protection, thermometers, and fluorescent light bulbs [11]. However, at elevated concentrations, it becomes detrimental to the organisms and environment.

As those pesticides and heavy metals have been asses to be carcinogenic, toxic, for human being and the environment through the accumulation processes [12], many electrochemical researchers are developing devices/sensors able to detect, analyze, and quantify them in aqueous and non-aqueous media [10, 12–18]. Moreover, the method of their preparation which is the first crucial step for electroanalysis has to be investigated properly, and it is one of the objectives of this chapter. Thus, sample preparation is the bottleneck for the effective and accurate analysis of trace pesticide and heavy metal residues.

The sample preparation methods in electrochemistry are purely dependent on the state (liquids or solids) and the type (organic or inorganic) of the sample to be investigated. Methods commonly used for the preparation of samples are dissolution, extraction, reaction with some chemical species, filtering, dilution, and emulsion [19]. Why do we need to prepare a sample? To isolate traces of analyte from a large quantity of complex matrices and eliminate matrix interferences as far as possible. This is achieved by successive sampling/homogenization, extraction, and cleaning.

In the ideal situation for a solid sample, in aqueous media, the sample is first dissolved with water/buffer/acid and diluted with SE before analysis in both cases pesticide and heavy metal, while the liquid sample has to be dissolved only with SE. In non-aqueous media, the solid sample is dissolved and diluted with the same solvent, while the liquid sample has to be diluted with a solvent playing the role of SE.

In real-life situation, the solid/liquid sample has to be treated to eliminate the effect of organic substances, and the extraction and filtration step will be required. Therefore, different methods of pretreating samples are widely used. In this chapter, the steps involved in preparing samples (pesticides and heavy metals) for electrochemical analysis will be described in detail with the difficulties and issues, prior to a brief discussion on the electrochemical methods of analysis of pesticides and heavy metals.

2. Brief review of electrochemical methods of analysis of pesticides and heavy metals

The proper electrochemical analysis of pesticides, heavy metals, dyes, food additives, etc. has all as the first step in the preparation of the sample prior to the sampling. Once this is done, the second task to be performed is to determine the electroactivity domain/potential of the analyte present in the electrochemical cell. This is done by CV, which would also give information about the electron transfer process. After the redox potential of the target identify, the next thing to do depending on the objective to have good sensitivity is to apply a pulse voltammetry technique like DPV, SWV, CA, NPV, LSV, ASV or CSV, etc. Moreover, in contrast to the electroanalysis of pesticides where the two steps are evolved, heavy metals which have known oxidation potential most of the time, during their electroanalysis, the experimenter does not use CV, but directly the pulse method, especially the LSV and ASV [12]. Therefore, for the analysis of pesticides and heavy metals, a brief description of CV and ASV is given below.

2.1 Cyclic voltammetry

CV is a very efficient electrochemical technique that enables the reactivity of electroactive substances to be studied quickly, easily, and affordably. It is generally used for the study of a new type of compound [20]. It makes it possible to determine oxidation or reduction potentials, but also to understand the reaction mechanisms involved in the reaction medium [21]. This technique is based on the change of potential scanned forward and backward between two potential limits (initial E_i and final E_f) while the current that circulates through the circuit is recorded in the function of the potential for the oxidation and reduction of the compound [22]. During scanning, the current intensities are measured in the anode (I_a) and in the cathode (I_c) and the curve obtained represents the variation of the current resulting from the various electrochemical processes as shown by the reversible process given in **Figure 1a**. Moreover, depending on the kinetic of the electrode, electrochemical reactions can be described as reversible (**Figure 1a**), quasi-reversible (**Figure 1b**), and irreversible systems (**Figure 1c**).

For a system to belong to one of the listed systems, apart from the shape difference shown in **Figure 1**, for a transfer of one electron at 25°C, the peak-to-peak separation ($\Delta E_p = E_i - E_f$) must approach 59 mV for a reversible system. A value above 60 mV suggests that the electron transfer is either quasi-irreversible or irreversible. Another difference but not all is based on the peak ratio, that is for reversible and quasi-reversible systems, $Ia/Ic = 1$ while for irreversible systems $I_a/I_c \neq 1$. A practical example in the case of pesticides is shown in **Figure 2**. The multisweep cyclic voltammograms of methyl parathion (MP) pesticide shown in **Figure 2a** shows two redox systems reversible ($E_{a1} = 0.03$ V and $E_{c1} = 0.00$ V, for oxidation and reduction respectively) and irreversible ($E_{c2} = -0.60$ V, for reduction peak) [23], while **Figure 2b** shows the multisweep cyclic voltammograms of diuron pesticide where a single oxidation peak ($E_a = 1.22$ V) for an irreversible process is observed [14].

2.2 Stripping methods

Stripping voltammetry is a sensitive electroanalytical technique used for the determination of trace amounts of metals and pesticides in solution. This technique comprises two successive steps. First of all, an accumulation of material (deposit) is carried out on (or in) the electrode by electrolysis at a controlled potential carried out under well-defined conditions (agitation, duration of electrolysis, temperature, etc.). Then, after a short period of rest, re-dissolution is carried out by imposing a suitable potential sweep, while the intensity of the current passing through the electrode is recorded as a function of time. There are several stripping techniques such as ASV, CSV, adsorptive stripping voltammetry (AdSV), and potentiometric stripping analysis (PSA) [24]. Moreover, the one generally used for the electroanalysis of heavy metals

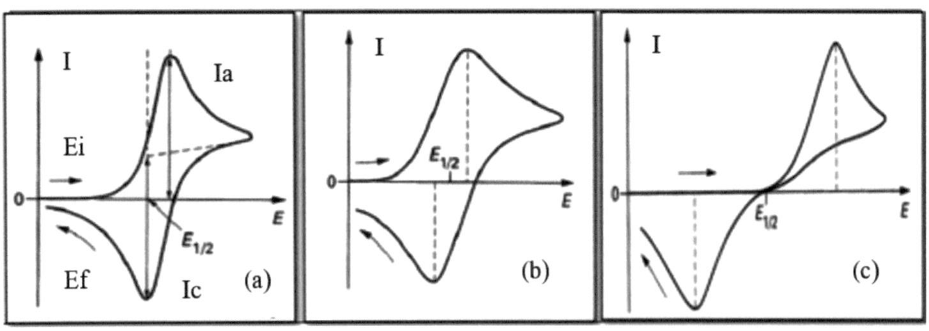

Figure 1.
Typical shape of cyclic voltammograms for different systems: reversible (a), quasi-reversible (b), and irreversible (c) [20]. (From [21]. Copyright (2014), with permission from Dunod.)

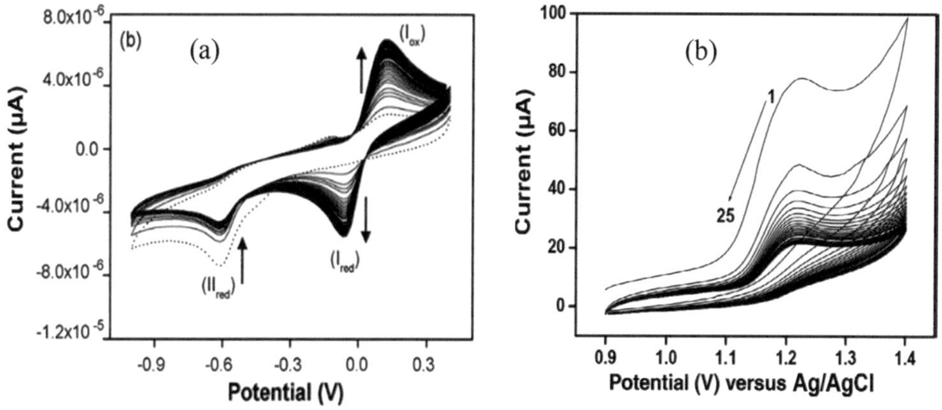

Figure 2.
(a) Multisweep cyclic voltammograms recorded in 41.5 μM MP in 0.1 mol l^{-1} acetate buffer (pH 5) on the GCE coated by the Ba-DO_4 organoclay film. Potential scan rate: 100 mV s^{-1}. (b) Multisweep cyclic voltammograms of diuron 5 × 10^{-4} M recorded on GCE/Sa (DODAB)/poly(ARS) in PBS, pH 1, scan rate = 50 mV s^{-1} [14, 23]. (Copyright (2022 and 2010), with permission from Wiley and Elsevier.)

is the ASV. The principle of ASV is based on electrolysis at a controlled potential. It consists of a reduction, the re-dissolution being carried out by an anodic sweep. By pre-electrolysis, in a stationary convective diffusion regime, at a sufficiently low fixed potential, the cathodic reduction of the metal cation(s) in solution (complexed or not) to metal deposited on the surface of the electrode is produced. This produces the accumulation of a fraction of the element initially diluted in the solution, which becomes "concentrated" in the form of a deposit. After a certain delay intended to return to a regime of pure natural diffusion (without agitation or convection), it is then possible to proceed with a potential sweep of the electrode toward the positive values to cause the re-oxidation of the metallic deposit and thus to cause a current signal corresponding to the pre-concentrated metallic element. The duration of the necessary pre-electrolysis is usually of the order of 2–10 min (1 min is then necessary to ensure that the solution has come to rest). A practical example of the ASV analysis and control of traces of Pb in drinking water is given in **Figure 3**. **Figure 3A** shows the influence of modification of the electrode for the ASV determination of Pb^{2+} [25], while **Figure 3B** shows the influence of the concentration and the calibration curve for the determination of the limit of detection (LOD) of Pb^{2+} using the ASV method [26].

3. Sample preparation for electrochemical analysis of pesticides

Pesticides are represented as substances or mixtures designated for destroying any group of pests such as insects and plants [27]. Pesticides have led to the improvement of food production, to secure the demands of the increasing human population reason while it is one of the largely used amounts in the world. Moreover, under the degradation process of pesticides which can be biotic or abiotic, and under the accumulation processes, pesticides lead to the formation of residual products that are often even more toxic than the original compounds [3]. Therefore, significant impacts on controlling pests are beneficial to prevent hazardous diseases in agriculture crops and the environment. Electrochemistry is one of the most used techniques for the analysis of pesticides in these decays. For that to be performed, the preparation of pesticides

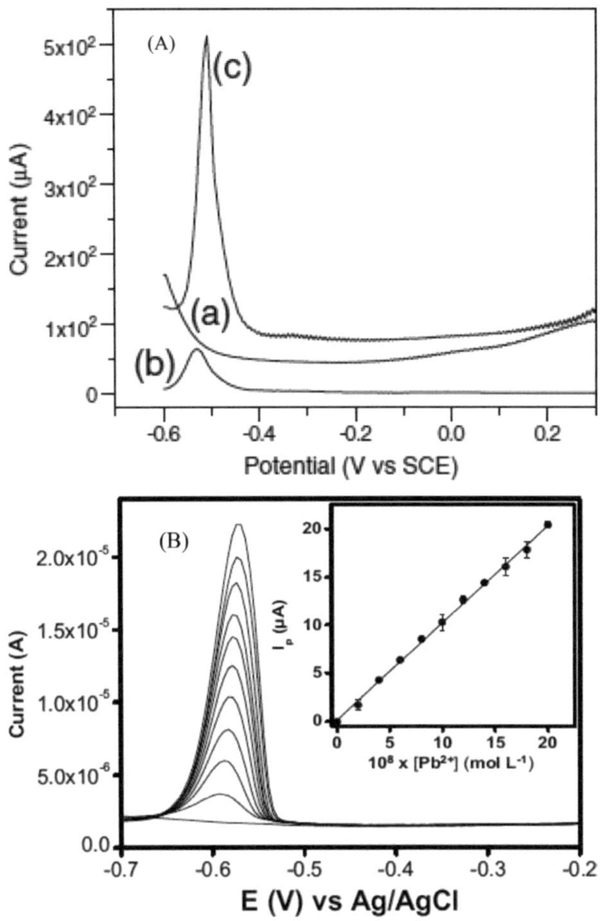

Figure 3.
(A) Influence of the modification of the clay mineral on the electrochemical response: curves recorded in 2 MHNO_3 (a) at OP/CPE; (b) at Ba-Na^+/CPE; and (c) at OP (40 mM)-Ba/CPE after 2 min accumulation in 1.5×10^{-5} M Pb (II) at pH 6. Detection by square wave ASV after 60 s electrolysis at −0.8 V. (B) Differential Pulse ASV signal recorded at GCE/$NHAP_{P0.5}$-CA-β-CD under optimized conditions, after 10 min accumulation in an aqueous medium containing Pb^{2+} in the range from 2×10^{-8} to 2×10^{-7} M. Other experimental conditions are as follows: 0 s equilibration time, 5 mV potential step and 100 mV amplitude. The inset shows the corresponding calibration curve [25, 26]. (Copyright (2011 and 2021), with permission from Elsevier and Springer.)

(solid or liquid) has to be done. We illustrate in the following paragraph extraction methods which are the technique for the preparation of pesticides in the environment and foodstuffs.

Traditional sample extraction methods, in particular liquid–liquid extraction (LLE), have been widely used for the analysis of pesticide residues. For the electroanalysis of pesticides in food, new methods have emerged, such as Quick, Easy, Cheap, Efficient, Rugged, and Safe (QuEChERS), solid-phase extraction (SPE), supercritical fluid extraction (SFE), solid-phase micro-extraction (SPME), liquid-phase micro-extraction (LPME), molecularly impregnated polymers (MIP), pressurized liquid extraction (PLE), microwave-assisted extraction (MAE), matrix solid-phase dispersion (MSPD), and ultrasound-assisted extraction (UAE). The most used one for sample preparation in electrochemistry are LLE, SPE, and QuEChERS [8].

3.1 Liquid–liquid extraction

LLE is known as the oldest extraction method which is most commonly used up to today. Before the extraction takes place in LLE, solid samples are to be processed into fine, homogeneous particles through mechanical grinding, mixing, rolling, stirring, chopping, crushing, maceration, chopping, pressing, or pulverization, among other methods. Now, whether they are solids or liquids, they are extracted several times with an immiscible organic solvent, followed by centrifugation, concentration and/or purification, and used for the final analysis. A practical situation was described in ref. [28] where they have immobilized the enzyme acetylcholinesterase on a composite of polyaniline/carbon nanotube, to form a biosensor used for the electrochemical analysis of carbamate pesticides in fruit and vegetables. The work reported in ref. [29] highlights the development of an improved method for extracting pesticide residues from food using acetonitrile (MeCN) with magnesium sulfate and chloroform. The method involves homogenizing a mass (10 g) of fruit, vegetable, and meat samples and placing them in a teflon centrifuge tube (V = 50 ml) containing 10 ml of MeCN, and the mixture is shaken vigorously in a vortex mixer for 1 min. Next, 4 g of $MgSO_4$ and 1.0 ml of chloroform were added, and shaken for 1 min. Finally, the extract was centrifuged at 4000 rpm for 10 min to separate the phases. In ref. [30], the authors successively used centrifugation, low-temperature purification at −20°C for 6 h, and dispersive solid phase extraction (d-SPE) with Florisil as cleaning steps to perform the LLE. The efficiency of the extraction of analyte in LLE depends mainly on the distribution/partition coefficient equilibrium between the donor and acceptor phases, which requires the polarities of the extraction solvents and analytes to be matched according to the principle of similarity. The miscibility of MeCN with water has been shown to be effective in extracting polar and/or non-polar pesticide residues with low amounts of matrix co-extracts, as they can be separated from the aqueous phase by the salting-out effect [8]. It is necessary to clean the extracts to reduce interferences after the initial LLE. LLE is a classic method for routine sample preparation due to its simplicity, robustness, and efficiency. However, it remains difficult to implement, with high consumption of solvents and time. It is, therefore, important to choose simple and rapid sample preparation methods for the analysis of pesticide residues in foods.

3.2 Solid phase extraction

First introduced in 1970 [31], the principle involves passing extracts through a cartridge. These extracts (analytes) are adsorbed onto solid phase materials (solid sorbents), which have been previously conditioned and activated with water and/or an organic solvent before being used. The choice of an appropriate sorbent is, therefore, very crucial. Interferences are eliminated by pre-washing with organic solvents, while any analytes still retained on the adsorbents can be eluted with other organic solvents so that clean extracts should be obtained. As a trump card, SPE requires small volumes of solvent and standard experimental equipment. It also offers a simple experimental procedure and rapid sample throughput, which is why it is used as an alternative to LLE both for sample preparation and for the clean-up and enrichment of organic compounds in water samples. The work reported in ref. [32] highlights the applications of SPE for sample clean-up in the analysis of pesticide residues in the food matrix. In this work, agricultural samples (chestnut, shallot, ginger, and tea) were first extracted by freeze-drying for approximately

24 h and ground to a homogeneous powder under liquid nitrogen protection. The homogenized samples (0.5 g of tea sample) were placed in a polypropylene centrifuge tube containing 1 ml of water was added and soaked for approximately 30 min. Next, 0.4 ml of D4-labeled imidacloprid and 10 ml of acetonitrile were added in the centrifuge tube and the sample was shaken vigorously for 30 s and then vortexed for 1 min. This extraction procedure was repeated three times with 10 ml acetonitrile and the solvent extracts were then combined. After separation of the organic layer, it was dried under anhydrous sodium sulfate for 30 min, then concentrated to approximately 4 ml under vacuum at 50°C. Activated carbon and Oasis HLB SPE cartridges were used for SPE. The pesticides retained by the activated carbons were eluted with methanol and the eluent was collected in a test tube, then reduced to dryness by rotary evaporation under vacuum on a water bath at 50°C. The residue was reconstituted in 2 ml of mobile phase and the final solution was filtered through a 0.22 m PTFE syringe filter.

The SPE procedure typically involves the selective retention of target analytes on an adsorbent packaged in a disposable mini-extraction column. Adsorbents such as C18, activated carbon, and multiwall carbon nanotubes have been developed for this purpose [33]. SPE is simpler, more practical, less solvent-intensive, and easier to automate than LLE, and it can effectively avoid the emulsion formation often encountered in LLE. However, it is difficult to quickly select suitable adsorbents and elution solvents for the analysis of multi-residue pesticides due to the complexity of their physicochemical properties. Secondly, commercial SPE cartridges are not recyclable, which considerably increases the experimental budget. Recently, a rapid clean-up step based on d-SPE was developed based on the principle of the QuEChERS method [8, 34].

3.3 Quick, easy, cheap, efficient, rugged, and safe

QuEChERS, which stands for quick (or fast), easy, cheap, efficient, robust, and safe, is a very flexible approach that can be modified depending on the properties of the target analyte, the composition of the matrix, and the facilities equipment available in the laboratory. The model allows high recoveries to be achieved for many pesticides in many matrices, even if different ratios, types of samples, solvents, salts, and sorbents are used for the modifications. In the analysis of multi-class and multi-residue pesticides, the sample preparation method inherently requires a large analytical scope, making it impossible to achieve a high degree of clean-up without reducing recoveries for some pesticides. Moreover, higher clean-up can be achieved by using different sorbents in d-SPE if the application has a reduced analytical scope [35].

Introduced in 2003 [36], the QuEChERS method is based on microscale extraction using MeCN, water absorption and liquid–liquid separation using $MgSO_4$ and NaCl, and the d-SPE clean-up step using a primary-secondary amine (PSA) adsorbent, as shown in **Figure 4**. By combining conventional extraction, isolation and clean-up procedures into a single step, this method avoids mixing, filtration, high-volume solvent transfer, evaporation, and condensation. It has become very popular in the preparation of samples for the analysis of pesticide residues in the food matrix. In the work reported in ref. [37], detection and quantification of cypermethrin with the enzyme glutathione S-transferase which catalyzed the reaction between glutathione and 1-chloro-2,4-dinitrobenzene in the presence of several solvents such as methanol and ethanol. The QuEChERS method was used to validate their study by fortifying

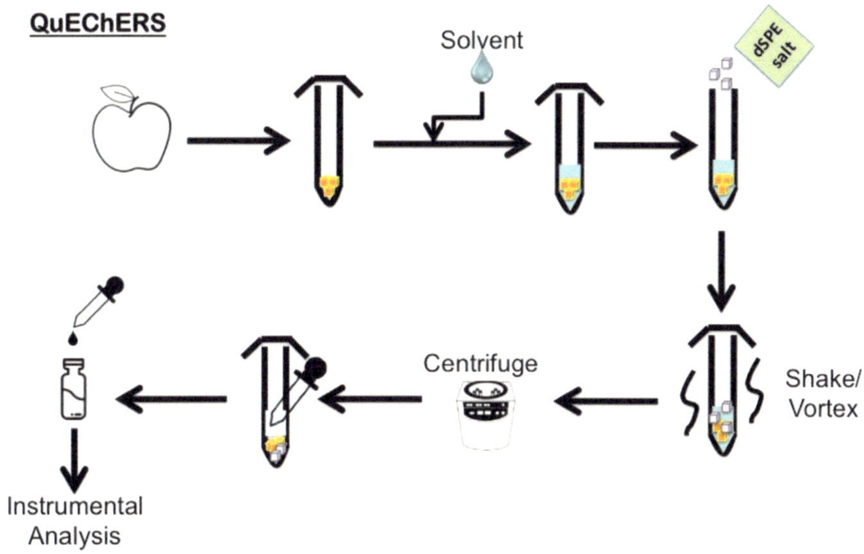

Figure 4.
Schematic diagram of the common procedure of QuEChERS [8]. (Copyright (2012), with permission from Wiley-VCH GmbH.)

tomato samples with a known quantity of cypermethrin. Among the solvent used in this QuEChERS procedure which are methanol, acetone, and MeCN. MeCN has been chosen as the best solvent not only because it helps to achieve higher recovery rates and less interference than other solvents such as acetone and methanol, but also to offer a slightly better LOD and relative standard deviation (RSD) than acetone. Compared with the classic LLE method, the QuEChERS method offers the advantage of a simpler, less time-consuming procedure and lower consumption of organic solvents. Since it considerably simplifies the extraction and cleaning stage during sample preparation and provides reliable quantitative results, QuEChERS has a bright future in the field of chemical analysis. It has a bright future in the analysis of pesticide residues in foods.

The general steps for the preparation of a sample for the electroanalysis of pesticides are illustrated **Figure 5** [38].

4. Sample preparation for electroanalysis of heavy metals

The daily applications of heavy metals in the manufacturing of several varieties of equipment used by human beings with the consequence of releasing some trace particles which can accumulate in the environment and pollute the foods and drinking water has become a serious problem to be solved. Heavy metal pollution has therefore become a priority, as they can affect biological functions or become accumulated in organs, causing illnesses such as cancer. These pollutants are released into soil, water, and the food chain as a result of a number of anthropogenic factors. This section provides a clear overview of the different sample preparation methods developed for electrochemical analysis of samples, which can be classified into liquids and solids, or organic and inorganic.

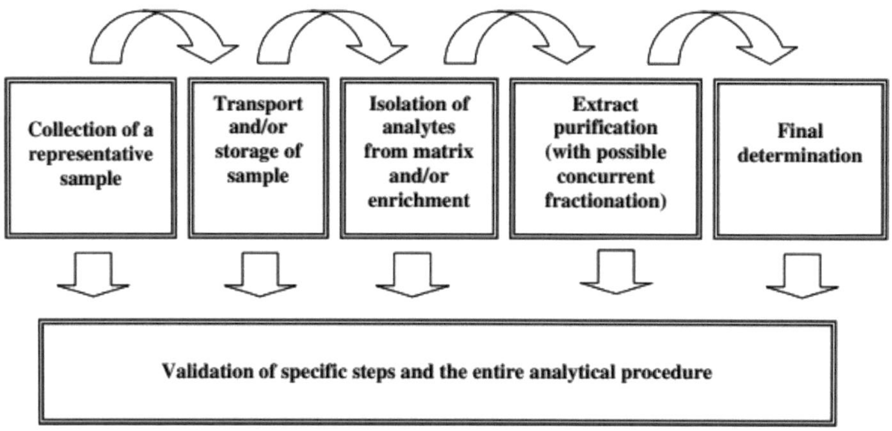

Figure 5.
Main steps of sample preparation for electroanalysis of pesticides [38]. (Copyright (2012), with permission from Elsevier.)

4.1 Types of samples

To analyze heavy metals in the environment and foods, samples can be liquids, solids, or organic and inorganic like in the case of pesticides. Sample preparation still needs to be optimized to meet the same requirements as the analytical procedures needed to determine the analytes [39]. Liquid samples may be in aqueous solution (e.g., natural water, beverages, and biological fluids), but also in organic form (oils and fuels). Solid samples may be predominantly organic in nature (e.g., foods, plants, and biological tissues) or predominantly inorganic (e.g., soils and sediments). Once samples have been collected, they can be analyzed by electroanalytical re-dissolution techniques either directly or after dilution.

4.2 Sample pre-treatment

Prior to the actual processing step, sample pre-treatment can be carried out depending on the solid or liquid nature of the sample. For solid samples or solutions made directly from wet-washed materials or dry-washed and dissolved materials [40], the amount (size) of material processed can be reduced while maintaining the representativeness of the subsample by mechanical grinding, by abrasion in a cyclone mill or by crushing in a ball mill [40]. In order to eliminate or reduce physical interferences, certain techniques are also used, such as sample dilution, mapping of the template and blank to the sample, using internal standards, chemical separation, and the method of standard addition [41]. Dilution is carried out according to the concentration of the analyte and/or the interference present in the matrix.

4.3 Sample preparation

The metals contained in the samples are determined by several analytical methods. Metals such as copper in brass or iron in steel can easily be analyzed by methods such as gravimetric chemical methods. However, many metal determinations involve smaller quantities, or traces, which are determined by various methods like spectroscopic, chromatographic, and electrochemistry. Sample preparation for metal determination has several objectives, which vary according to the type of sample and the requirements of the analysis. The main objective of sample treatment is to ensure the release of metals from the organic matrix and to remove the sample matrix, which may interfere with electrochemical detection. Sample preparation techniques such as dry ashing, wet digestion, and extraction methods are frequently reported.

4.3.1 Dry ashing

Dry ashing was initially used for sample preparation due to its easy operation and effective destruction of organic materials [42]. Organic compounds are ignited and oxidized by air at elevated temperatures (~500°C) and atmospheric pressure. The carbonaceous residues (ashes) are dissolved in diluted acid solutions (HNO_3 or HCl solutions). For stripping analysis, a substantial dilution (0.5 g per 100 ml) may be necessary to circumvent the interference from sample ash on the electrochemical detection. For example, in ref. [43] dry ashing was used to prepare vegetable samples for Zn, Cd, and Pb determination. The samples were dry-ashed (in a muffle oven and the temperature was slowly increased up to 500°C) until they yielded white ash, which was then dissolved by a concentrated nitric acid/perchloric acid mixture (4:1), heating gently on a hot plate until almost dry. Despite the simplicity of this process, it cannot be used in some important applications where high temperatures and long decomposition times are used since they can lose in whole or in part [44]. Some of the volatile elements likely to be lost are As, B, Cd, Cr, Cu, Fe, Pb, Hg, Ni, P, V, and Zn [45]. Microwave ovens can incinerate samples in around 20 min, significantly reducing analysis time by 97% compared to normal dry incineration. Reaching temperatures of up to 1200°C, microwave ovens typically digest under high pressure in closed containers heated by microwave energy, breaking down the sample faster while minimizing external contamination and volatile chemical losses [41]. These ovens take more samples than traditional ovens [45]. The advantages of conventional dry incineration are that it is a safe and affordable method that requires fewer samples than other methods, does not require additional acids or other reagents or blank subtractions, has a high sample recovery rate and once ignition has started minimal attention is required [45].

To reduce sample preparation time and to work at lower temperatures, wet digestion processes have been an interesting alternative to dry ashing methods.

4.3.2 Wet digestion processes

In wet digestion, primarily for the preparation of mineral analyses, in ref. [45], they have used strong oxidizing acids (mixtures of nitric and perchloric acids (HNO_3–$HClO_4$), sulfuric acid (HNO_3–H_2SO_4), hydrochloric acid (HNO_3–HCl), and hydrogen peroxide (HNO_3–H_2O_2)) that are added to the sample and heated to facilitate the wet digestion procedures by decomposing the organic components of the sample. The maximum temperatures reached are between 122°C and 338°C at ambient pressure and are limited by the acid or acid mixture used [46]. A disadvantage is the contamination that can occur during digestion from reagents, vessel materials, and the environment. However, wet microwave incineration can be carried out safely in an open or closed vessel microwave system. Acids can be heated above their boiling point due to the ability of closed vessels to withstand higher pressures. This technique allows more samples to be processed at one time than conventional techniques, thus increasing throughput. Normal digestions take less than 30 min. Compared to conventional methods, microwave digestion procedures are preferred since they have short digestion time, low acid usage, and high extraction efficiency [47]. An important parameter to be evaluated when stripping techniques are utilized is the residual organic matter from the digestion process, which may interfere with stripping voltammetric determinations. In the work reported in ref. [48], wet acid digestion was used to destroy organic matter in milk samples and fermented milk products for the determination of cadmium, lead, and copper. The digestion was carried out in quartz Kjeldahl flasks, using a mixture of nitric and perchloric acids (1:1).

4.3.3 Extraction methods

The most commonly used extraction methods from solid or immiscible liquid samples are in the aqueous phase containing acids or oxidants, with the main advantage that the operation is simple, does not require drastic pressure and temperature, and does not require sophisticated equipment. For example, cadmium and lead were extracted from the soil sample, just by heating 1 g of the sample in an oven at 80°C for 1 h. The soil sample was then placed in an extraction tube with 40 ml of 0.1 mol l^{-1} acetic acid [49]. Extractions can be improved if external energy is introduced, such as increasing the temperature of the sample-extractor medium. For example, MAE of metals from soils has been reported for the determination of metals by stripping techniques [50]. In addition, ultrasound has also been used for sample preparation, in particular, to facilitate extraction methods. An extraction performance close to 100% has been achieved using hot concentrated HCl (50–90°C) [51] for the determination of metals in petroleum-based oil products. Eight types of retail packaged milk were treated (extraction process) by the ultrasonic-assisted acid digestion method for the electrochemical determination of cadmium and lead [52]. Some extractions are time-consuming and when organic samples are processed, the residual organic matter in the aqueous extractor is not negligible and therefore affects the electrochemical determinations. It is therefore sometimes necessary to use carbon-retaining columns prior to stripping determinations of metals in the aqueous extractor [51]. Removal of these species can also be done by organic solvent extraction [53].

4.4 Choice of sample preparation methods

Many environmental sample preparation procedures focusing on the determination of metals by stripping techniques have been highlighted in the literature [19]. **Figure 6** summarizes the sample preparation methods for the determination of metals in environmental, biological, and food samples by electrochemistry. Acid digestion methods have been favored when analyzing crude oil and petroleum-based products due to their extremely high organic content. However, the complete decomposition of these matrices requires extreme conditions (high temperature and pressure in the presence of concentrated acids and oxidants) that can be avoided by using ultrasonic and room-temperature extraction methods. Extraction methods are the predominant option for solid environmental samples.

5. Conclusion

The methods described in this perspective chapter concern the determination of pesticides and heavy metals. However, heavy metals are present in different forms in

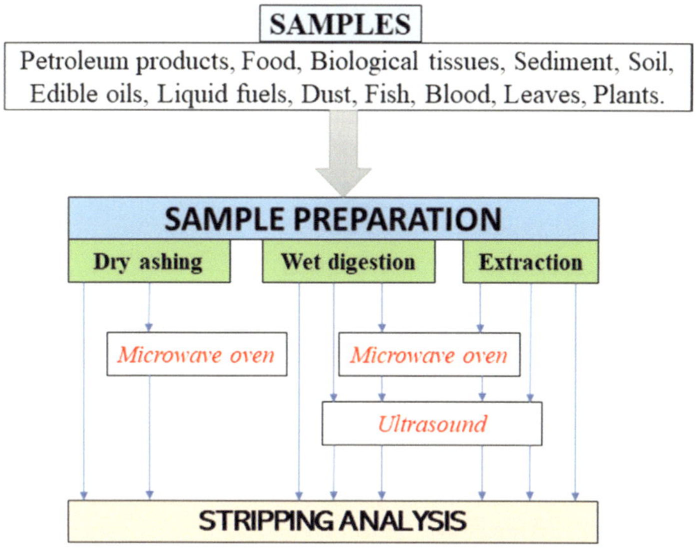

Figure 6.
Summary of sample preparation techniques for the re-dissolution determination of metals in environmental and food samples.

environmental and food samples. As a result, the use of highly sensitive electrochemical techniques for the determination of pesticides and heavy metals in environmental and food samples is highly dependent on the sample preparation step, especially when analyzing solid-organic material. Technical developments have made it possible to achieve efficient, fast, and inexpensive sample preparation procedures. For the determination of heavy metals, the combination of microwaves and ultrasound with acid digestion or extraction provides important advantages such as the reduction of reagents and samples (environmental, biological, and food) and the reduction of analysis time. The introduction of an ultrasonic horn into the electroanalytical system is a promising methodology to perform the analysis of complex samples in one step (sample preparation, electrode cleaning, and stripping analysis occurring simultaneously in the electrochemical cell). In the case of the analysis of pesticides, the most used technics for sample preparation in electrochemistry are LLE, SPE, and QuEChERS. Among them, QuEChERS was found to be a very versatile approach and serves as a model for modification according to the properties of the analyte, the composition of matrix, and the facilities technique on hand in the laboratory. The toxicity and the bioavailability of pesticides and heavy metals depend on their form. The main challenges in their determination are sample preparation, the presence of interferences, detection limits, and matrix effects. To date, pesticides and heavy metals have been determined in various types of samples, including environmental and food samples. Quality control measures such as blanking, addition of a standard, spike recovery, repeated analysis, and the use of reference materials remain crucial to ensure accurate results. In addition to pesticides and heavy metals, other xenobiotics such as drugs, dyes, and flame retardants can also be analyzed by electrochemical methods, with sample preparation almost identical to those described above. However, there are other, fairly sophisticated and robust analytical methods that allow these analytes to be analyzed efficiently. These include mass spectrometry (for organic compounds and pesticides) and inductively coupled plasma (for heavy metals).

Acknowledgements

The authors thank the Research Unit of Noxious Chemistry and Environment Engineering (RUNOCHEE) and the Yaoundé-Bielefeld Graduate School Natural Products with Antiparasite and Antibacterial Activity (YaBiNaPA) project for all their support and facilities.

Conflict of interest

The authors declare no conflict of interest.

Abbreviations

Ia	anodic current
ASV	anodic stripping voltammetry
AE	auxiliary electrode
Ic	cathodic current
CSV	cathodic stripping voltammetry
CA	chronoamperometry
CV	cyclic voltammetry
DPV	differential pulse voltammetry
EC	electrochemistry
LSV	linear sweeping voltammetry
LC	liquid chromatography
LPME	liquid phase micro-extraction
LLE	liquid–liquid extraction
MS	mass spectrometry
MSPD	matrix solid-phase dispersion
MP	methyl parathion
MAE	microwave-assisted extraction
MIPs	molecularly imprinted polymers
NPV	normal pulse voltammetry
PSA	potentiometric stripping analysis
PLE	pressurized-liquid extraction
RE	reference electrode
SPE	solid-phase extraction
SPME	solid-phase micro-extraction
SWV	square-wave voltammetry
SFE	supercritical-fluid extraction
SE	supporting electrolyte
UAE	ultrasound-assisted extraction
UV-Vis	ultra-violet
WE	working electrode

Author details

Ranil Clément Tonleu Temgoua[1,2*], Sherman Lesly Zambou Jiokeng[1], Kevin Yemele Tajeu[1], Gullit Deffo[1], Lionnel Averie Vomo[1], Fabiola Tonleu Dontsi[1], Evangeline Njanja[1] and Ignas Kenfack Tonlé[1]

1 Electrochemistry and Chemistry of Materials, Department of Chemistry, University of Dschang, Dschang, Cameroon

2 Department of Chemistry, University of Yaounde I, Higher Teacher Training College, Yaounde, Cameroon

*Address all correspondence to: raniltemgoua@yahoo.fr

References

[1] Rieger PH. Mechanisms of Electrode Processes. In: Electrochemistry. Dordrecht: Springer; 1994:247-314. DOI: 10.1007/978-94-011-0691-7_5

[2] Hibbert DB. Introduction to electrochemistry. In: Introduction to electrochemistry. Macmillan Physical Science Series. London: Palgrave; 1993. pp. 1-10. DOI: 10.1007/978-1-349-22721-1_1

[3] Temgoua RCT, Bussy U, Alvarez-Dorta D, Galland N, Hémez j, Tonlé IK, Boujtita M. Simulation of the environmental degradation of diuron (herbicide) using electrochemistry coupled to high resolution mass spectrometry. Electrochimica Acta. 2020;**352**:136352. DOI: 10.1016/j.electacta.2020.136485

[4] Deffo G, Tene TFN, Dongmo LM, Jiokeng SLZ, Temgoua RCT. Differential pulse and square-wave voltammetry as sensitive methods for electroanalysis applications. Reference Module in Chemistry, Molecular Sciences and Chemical Engineering. Elsevier B.V., 2023:1-9. DOI: 10.1016/B978-0-323-85669-0.00040-4

[5] Climent V, Feliu JM. Cyclic voltammetry. Molecular Sciences and Chemical Engineering. 2015;**1**:28

[6] Bard AJ, Faulkner LR. Electrochemical Methods: Fundamentals and Applications. Paris, France: John Wiley and Sons Inc.; 2001

[7] Carter A. How pesticides get into water and proposed reduction measures. Pesticides Outlook. 2000;**149**:157-111

[8] Zhang L, Liu S, Cui X, Pan C, Zhang A, Chen F. A review of sample preparation methods for the pesticide residue analysis in foods. Central European Journal of Chemistry. 2012;**9**(00):925-910. DOI: 10.2478/s11532-012-0034-1

[9] Appenroth KJ. Definition of "Heavy Metals" and their role in biological systems. Soil Heavy Metals. 2009;**19**:29. DOI: 10.1007/978-3-642-02436-8_2

[10] Guenang LS, Dongmo LM, Jiokeng SLZ, Kamdem AT, Doungmo G, Tonlé IK, et al. Montmorillonite clay-modified disposable ink-jet-printed graphene electrode as a sensitive voltammetric sensor for the determination of cadmium(II) and lead(II). SN Applied Sciences. 2020;**2**:476. DOI: 10.1007/s42452-020-2283-5

[11] Nordberg GF, Fowler BA, Nordberg M. In: Handbook on the Toxicology of Metals. 4th ed. Elsevier B.V.; 2015:765-786

[12] Jiokeng SLZ, Dongmo LM, Ymélé E, Ngameni E, Tonlé IK. Sensitive stripping voltammetry detection of Pb (II) at a glassy carbon electrode modified with an amino-functionalized attapulgite. Sensors and Actuators B. 2017;**1027**:1034-1242

[13] Noori JS, Mortensen J, Geto A. Recent development on the electrochemical detection of selected pesticides: a focused review. Sensors. 2020;**20**(8):2221-2220. DOI: 10.3390/s20082221

[14] Deffo G, Temgoua RCT, Tajeu KY, Njanja E, Doungmo G, Tonle IK, et al. Signal amplification by electropolymerization of alizarin red S for improved diuron detection at organosmectite modified glassy carbon electrode. Journal of the Chinese Chemical Society. 2022;**69**:349-358

[15] Mbokou FS, Karanga Y, Bako YFR, Tapsoba I, Sakande F, Koulibaly B, et al. Electrochemical determination of diuron in soil using a nanocrystalline cellulose modified carbon paste electrode. International Journal of Electrochemical Science. 2021;**16**:210552. DOI: 10.20964/2021.05.31

[16] Temgoua RCT, Bussy U, Alvarez-Dorta D, Galland N, Hémez J, Thobie-Gautier C, Tonlé IK, Boujtita M. Using electrochemistry coupled to high resolution mass spectrometry for the simulation of the environmental degradation of the recalcitrant fungicide carbendazim. Talanta. 2021;**221**:121448

[17] Rajagopalan V, Venkataraman S, Rajendran DS, Kumar VV, Kumar VV, Rangasamy G. Acetylcholinesterase biosensors for electrochemical detection of neurotoxic pesticides and acetylcholine neurotransmitter: a literature review. Environmental Research. 2023;**227**:115724

[18] Meng R, Zhu Q, Long T, He X, Luo Z, Gu R, Wang W, Xiang P. The innovative and accurate detection of heavy metals in foods: a critical review on electrochemical sensors. Food Control. 2023;**150**:109743

[19] Munoz RAA, Almeida ES, Angnes L. Sample preparation techniques for the electrochemical determination of metals in environmental and food samples. Reference Module in Chemistry, Molecular Sciences and Chemical Engineering. 2013;**1**:11. DOI: 10.1016/B978-0-12-409547-2.05750-4

[20] Zoski CG. Electrochemical methods. In: Handbook of Electrochemistry. 1st ed. Vol. 836. Amsterdam, Holland: Elsevier; 2007. pp. 848-934. ISBN 10 0-444-51958-0

[21] Miomandre F, Sadki S, Audebert P, Méallet-Renault R. Aspects thermodynamiques de la reaction électrochimique, dans: Electrochimie des concepts aux applications. 3ème Ed. Vol. 261. Paris, France: Dunod; 2014. pp. 280-409. ISBN 978-2-10-071234-2

[22] Wang J. Controlled-potential techniques. In: Analytical Electrochemistry. 3rd ed. Vol. 67. New Jersey, USA: Wiley-VCH; 2006. pp. 114-250. ISBN 100-471-67879-1

[23] Tcheumi HL, Tonle IK, Ngameni E, Walcarius A. Electrochemical analysis of methylparathion pesticide by a gemini surfactant-intercalated clay-modified electrode. Talanta. 2010;**972**:979-981. DOI: 10.1016/j.talanta.2010.01.049

[24] Farghaly OA, Abdel Hameed RS, Abu-Nawwas A-AH. Analytical application using modern electrochemical techniques. International Journal of Electrochemical Science. 2014;**3287**:3318-3319

[25] Bouwe RGB, Tonle IK, Letaief S, Ngameni E, Detellier C. Structural characterisation of 1,10-phenanthroline–montmorillonite intercalation compounds and their application as low-cost electrochemical sensors for Pb (II) detection at the sub-nanomolar level. Applied Clay Science. 2011;**258**:265-252. DOI: 10.1016/j.clay.2011.02.028

[26] Tchoffo R, Ngassa GBP, Doungmo G, Kamdem AT, Tonlé IK, Ngameni E. Surface functionalization of natural hydroxyapatite by polymerization of β-cyclodextrin: application as electrode material for the electrochemical detection of Pb (II). Environmental Science and Pollution Research. 2021;**1**:14. DOI: 10.1007/s11356-021-15578-8

[27] Mrabet KE, Charlet P, Lalère B. Les pesticides. Laboratoire National de Métrologie et d’Essais, Paris, France. 2008;**1**:18

[28] Cesarino I, Moraes FC, Lanza MRV, Machado SAS. Electrochemical detection of carbamate pesticides in fruit and vegetables with a biosensor based on acetylcholinesterase immobilised on a composite of polyaniline-carbon nanotubes. Food Chemistry. 2012;**873**:879-135. DOI: 10.1016/j.foodchem.2012.04.147

[29] Liu G, Rong L, Guo B, Zhang M, Li S, Wu Q, et al. Development of an improved method to extract pesticide residues in foods using acetontrile with magnesium sulfate and chloroform. Journal of Chromatography A. 2011;**1218**:1429-1436. DOI: 10.1016/j.chroma.2011.01.041

[30] Nguyen TD, Lee MH, Lee GH. Rapid determination of 95 pesticides in soybean oil using liquid–liquid extraction followed by centrifugation, freezing and dispersive solid phase extraction as cleanup steps and gas chromatography with mass spectrometric detection. Microchemical Journal. 2010;**95**:113-119. DOI: 10.1016/j.microc.2009.11.009

[31] Sabik H, Jeannot R, Rondeau B. Multiresidue methods using solid-phase extraction techniques for monitoring priority pesticides, including triazines and degradation products, in ground and surface waters. Journal of Chromatography A. 2000;**885**:217-236. DOI: 10.1016/S0021-9673(99)01084-5

[32] Xie W, Han C, Qian Y, Ding H, Chen X, Xi J. Determination of neonicotinoid pesticides residues in agricultural samples by solid-phase extraction combined with liquid chromatography–tandem mass spectrometry. Journal of Chromatography A. 2011;**1218**:4426-4433. DOI: 10.1016/j.chroma.2011.05.026

[33] Ravelo-Pérez LM, Hernández-Borges J, Rodríguez-Delgado M. Multi-walled carbon nanotubes as efficient solid-phase extraction materials of organophosphorus pesticides from apple, grape, orange and pineapple fruit juices. Journal of Chromatography A. 2008;**33**:33-42. DOI: 10.1016/j.chroma.2008.09.084

[34] Walorczyk S, Drożdżyński D, Gnusowski B. Multiresidue determination of 160 pesticides in wines employing mixed-mode dispersive-solid phase extraction and gas chromatography–tandem mass spectrometry. Talanta. 2011;**85**:1856-1870. DOI: 10.1016/j.talanta.2011.07.029

[35] Lehotay SJ, Son KA, Kwon H, Koesukwiwat U, Fu W, Mastovska K, et al. Comparison of QuEChERS sample preparation methods for the analysis of pesticide residues in fruits and vegetables. Journal of Chromatography A. 2010;**1217**:2548-2560. DOI: 10.1016/j.chroma.2010.01.044

[36] Anastassiades M, Lehotay SJ, Stajnbaher D, Schenck FJ. Fast and easy multiresidue method employing acetonitrile extraction/partitioning and "Dispersive Solid-Phase Extraction" for the determination of pesticide residues in produce. Journal of AOAC International. 2003;**86**:412-431. DOI: 10.1093/jaoac/86.2.412

[37] Borah H, Dutta RR, Gogoi S, Puzari P. Influence of methanol, ethanol and cypermethrin on the glutathione S-tranferase catalyzed reaction of glutathione with 1-chloro-2,4-dinitrobenzene: a method for detection and quantification of cypermethrin. Electrochimica Acta. 2016;**205**:198-206. DOI: 10.1016/j.electacta.2016.03.199

[38] Beyer A, Biziuk M. Applications of sample preparation techniques in the analysis of pesticides and PCBs in food. Food Chemistry. 2008;**108**:669-680. DOI: 10.1016/j.foodchem.2007.11.024

[39] Zhang N, Shen K, Yang X, Li Z, Zhou T, Zhang Y, et al. Simultaneous determination of arsenic, cadmium and lead in plant foods by ICP-MS combined with automated focused infrared ashing and cold trap. Food Chemistry. 2018;**264**:462-470. DOI: 10.1016/j.foodchem.2018.05.058

[40] El Hosry L, Sok N, Richa R, Al Mashtoub L, Cayot P, Bou-Maroun E. Sample preparation and analytical techniques in the determination of trace elements in food: a review. Food. 2023;**12**:895. DOI: 10.3390/foods12040895

[41] Bolann BJ, Rahil-Khazen R, Henriksen H, Isrenn R, Ulvik RJ. Evaluation of methods for trace-element determination with emphasis on their usability in the clinical routine laboratory. Scandinavian Journal of Clinical and Laboratory Investigation. 2007;**67**:353-366. DOI: 10.1080/00365510601095281

[42] Mader P, Szakova J, Curdova E. Combination of classical dry ashing with stripping voltammetry in trace element analysis of biological materials: review of literature published after 1978. Talanta. 1996;**43**:521-534. DOI: 10.1016/0039-9140(95)01793-3

[43] Xu H, Zeng L, Huang D, Xian Y, Jin L. A Nafion-coated bismuth film electrode for the determination of heavy metals in vegetable using differential pulse anodic stripping voltammetry: an alternative to mercury-based electrodes. Food Chemistry. 2008;**109**:834-839. DOI: 10.1016/j.foodchem.2007.12.065

[44] Barnes RM, Júnior DS, Krug FJ. Introduction to sample preparation for trace element determination. Microwave-Assisted Sample Preparation for Trace Element Analysis. 2014. pp. 1-58. DOI: 10.1016/B978-0-444-59420-4.00001-5

[45] Harris GK, Marshall MR. Ash analysis. In: Nielsen SS, editor. Food Analysis. Food Science Text Series. Cham: Springer International Publishing; 2017. pp. 287-297. DOI: 10.1007/978-3-319-45776-5_16

[46] Hu Z, Qi L. 15.5-Sample digestion methods. In: Holland HD, Turekian KK, editors. Treatise on Geochemistry. 2nd ed. Elsevier; 2014. pp. 87-109. DOI: 10.1016/B978-0-08-095975-7.01406-6

[47] Güven D, Akinci G. Comparison of acid digestion techniques to determine heavy metals in sediment and soil samples. Gazi University Journal of Science. 2011;**24**:29-34

[48] Suturovic Z, Kravic S, Milanovic S, Durovic A, Brezo T. Determination of heavy metals in milk and fermented milk products by potentiometric stripping analysis with constant inverse current in the analytical step. Food Chemistry. 2014;**155**:120-125. DOI: 10.1016/j.foodchem.2014.01.030

[49] Ping J, Wu J, Ying Y, Wang M, Liu G, Zhang M. Evaluation of trace heavy metal levels in soil samples using an ionic liquid modified carbon paste electrode. Journal of Agricultural and Food Chemistry. 2011;**59**:4418-4423. DOI: 10.1021/jf200288e

[50] Nedeltcheva T, Atanassova M, Dimitrov J, Stanislavova L. Determination of mobile form contents of Zn, Cd, Pb and Cu in soil extracts by combined stripping voltammetry. Analytica Chimica Acta. 2005;**528**:143-146. DOI: 10.1016/j.aca.2004.10.036

[51] Coco FL, Monotti P, Rizzotti S, Ceccon L. Determination of lead in oil products by derivative potentiometric stripping analysis. Analytica Chimica Acta. 1999;**386**:41-46. DOI: 10.1016/S0003-2670(99)00002-1

[52] Ping J, Wang Y, Wu J, Ying Y. Development of an electrochemically reduced graphene oxide modified disposable bismuth film electrode and its application for stripping analysis of heavy metals in milk. Food Chemistry. 2014;**151**:65-71. DOI: 10.1016/j.foodchem.2013.11.026

[53] Safavi A, Maleki N, Shahbaazi HR. Indirect determination of hexavalent chromium ion in complex matrices by adsorptive stripping voltammetry at a mercury electrode. Talanta. 2006;**68**:1113-1119. DOI: 10.1016/j.talanta.2005.07.015

Chapter 3

Influence of Sewage Sludge and Heavy Fertilization on Nitrate Leaching in Soils: An Overview

Sidra Sohail, Muhammad Fraz Ali, Usman Zulfiqar, Saddam Hussain and Shaharyar Khosa

Abstract

Sewage sludge is now widely used for production of crops throughout the world. Utilizing the sewage sludge for crop production has various advantages such as nutrient recycling, reducing the need for mineral fertilizer, increasing organic matter amount in soils, and improving physical properties of soil. A large amount of sludge is generated during the sewage treatment process, and it is disposed off on land in the form of fertilizer or soil conditioner. In this, heavy metals are usually in greater concentration than in soil, which is one of the main problems which restricts its utilization as a fertilizer. Nitrate leaching and heavy fertilization cause serious problems for the groundwater and this issue of nitrate leaching is usually neglected. Continuous used of swage sludge cause contamination of soil and water and affects plant growth and soil microorganisms. In this chapter, we have focused on i) various factors which affects nitrate leaching in soil, it includes soil texture, rate of fertilization, season and climate, ii) strategies to reduce nitrate leaching and iii) nitrogen conversion mechanism in sewage sludge.

Keywords: sewage sludge, nutrient source, nitrogen conversion, nitrate leaching, heavy fertilization

1. Introduction

Nitrogen (N) is essential nutrient in agricultural system that influences various aspects of agroecosystems and other ecosystem services that are mainly depended on nitrogen availability in soil system [1, 2]. Nitrogen plays a cardinal role in the agroecosystem and energy balance of the Earth so any kind of change in nitrogen cycle will bring negative impacts on agroecosystem and health of other living beings [3]. Agroecosystems are extremely complex and have entangled webs of interactions between different plants, microorganisms, and soils, identifying critical components in crop production remain indescribable [4]. Consequently, understanding nitrogen fate in agroecosystems is really challenging.

Nitrogen plays a crucial role in life cycle of plants [5]. It is also involved in several critical processes, like growth, expansion of leaf area, production of biomass and

IntechOpen

increase in yield [6–8]. Agricultural crop production and yield mainly depends on N availability for protein synthesis [9], chlorophyll, photosynthesis and many other essential processes in plants [10]. Insufficient availability of N to the plants can hinder their growth and development. Nitrogen improves plant root growth, increase root volume, area of root, root diameter, total root length, root dry mass and subsequently enhance uptake of nutrients and dry mass production [11–13]. Nitrogen is lost from the agroecosystems in different forms like nitrate and nitrous oxide that contaminate aquatic ecosystems and also contribute to the climate change [14].

About 80% of the atmosphere is composed of N_2, that is relatively inert gas, it is further converted into other reactive forms of nitrogen like nitrate (NO_3^-) and ammonium (NH_4^+) through biological and industrial processes, to satisfy nitrogen requirement of crops [15, 16]. Nitrogen more important in the agricultural systems [17], nitrogen is added to sustain crop growth and improve yields that is a fundamental feature of modern crop management [18]. On the other side, there are some significant costs that are associated with the agricultural nitrogen additions. Nitrogen is mobile in nature [19], it is hard to sustain, even nitrogen that is conserved by the plants and taken away in harvesting eventually makes the way to return back to environment [20, 21]. Most of the nitrogen that is mobilized from the agricultural systems is highly reactive in nature; it is available in biologically active forms in soils and surface waters and chemically reactive forms in the atmosphere [22].

The main problem is excess of N that lead to N losses. This is difficult to supply precisely enough N to full fill plant physiological requirements while controlling reactive N fate to avoid losses into the environment [23, 24]. As agricultural system is responsible for food supply, and it leaks too much nitrate, causing degradation of aquatic ecosystems [25]. Nitrogen also loss in the form of gas from agroecosystem as ammonia, nitrogen oxides, and nitrous oxide that reduce quality of air and account for much contribution to the climate change [26, 27]. Therefore, there is an urgent need to make an agroecosystem that efficiently retain more nitrogen and that must remain productive as well as resilient to face changes of climatic conditions.

Nitrogen that is present in mineral forms are more prone to losses i.e. ammonia volatilization (loss of NH_3 from soil surface), denitrification and/or losses of nitrogen gasses (dinitrogen gas and nitrous oxide) and leaching (nitrate) [28]. Loss of nitrogen via leaching mainly occur in the nitrate form but sometimes NH_4^+ leaching may occur in light texture soils [29]. There are some other factors which are also involved in nitrate leaching. Irrigation amount and its distribution is responsible for nitrate leaching below plant root zone in the soil profile [30]. Soil structure may also affect movement of nitrate within the soil profile [31]. Sandy soil has poor soil structure so chances of leaching are more in sandy soils as competed to clayey soils having a good soil structure [32]. Additionally, macrofauna in the soil is also responsible for this phenomena, movement of macrofauna and growth of root allow speedy movement of nitrate in the soil profile [33].

There are many strategies to overcome nitrogen losses from various sources. Nitrate leaching risk is primarily reduced by decreasing inorganic fertilizers application and changing irrigation methods [34]. There are some other approaches that proved to be helpful in reducing nitrate leaching i.e. growing grass cover, application if controlled-release nitrogen fertilizer, nitrification inhibitors, organic amendments, etc. In recent years, organic amendments are being used by the farmers as soil improver. A well-known organic amendment sewage sludge (SS) contain nitrogen in higher amounts hence SS can be applied instead of inorganic fertilizer to provide sufficient nitrogen that is required to plants for growth and development. Sewage sludge contains 60% of total nitrogen in organic form and rest of 40% in present in mineral form [35]. Mineral N is released during decomposition of organic matter that is further immobilized in organic forms [36]. Organic form of nitrogen is slowly released into the soil, hence it reduces risks of groundwater contamination with nitrates [37]. Applying of sewage sludge into the soil improves texture of soil and slow down nutrients release from the soil. Sewage sludge is very useful in reducing nitrate losses from soils as it improves soil NUE for a sustainable production of crops. Sewage sludge plays a vital role in building up soil OM, increase fertility of soils and improve soil retention of inorganic N-forms e.g. nitrate. Heavy fertilization is a common practice used by farmers to obtain higher yield. Indiscriminate application of inorganic fertilizers are polluting our natural resources. This chapter reviewed the measures and methods to reduce leaching of nitrate from the agricultural soils. This review brings further insight to the role of sewage sludge to reduce leaching of nitrate from the soil by improving soil attributes.

1.1 Excessive use of nitrogen fertilizers

Nitrogen is a deficit nutrient in most of the agricultural soils [38, 39] and it is recommended to use nitrogen mineral fertilizer as external inputs for improving crop growth and yield [40, 41]. Many leguminous crops and soil microorganisms took part to fix nitrogen available for plant uptake [42]. However, to meet the target yields this is inadequate, so farmers are dependent on synthetic nitrogen fertilizers. Fertilization with these chemical fertilizers in excessive amounts is harmful to the environment [43], it causes an imbalance of several ecosystems' functions and services [44]. For sustainable agriculture and the ecosystem, judicious application of fertilizers plays major role.

Nitrogen fertilizer management has a vital role in sustainability, by reducing its various losses and enhancing nutrient uptake [45]. juidicious application of nitrogenous fertilizers and management of irrigation water are important practices that are helpful in reduction of nitrate leaching into the groundwater and they also improve NUE. Under current circumstances, integrated nitrogen management and balanced fertilizers use is needed. Gu et al. [46] conducted lysimetric research to monitor the role of irrigation and fertilization in nitrate leaching, results concluded that heavy fertilization of N fertilizer (urea) with high levels of irrigation led the nitrate leaching. It was also reported that manure applications have less N leaching than the urea application. The higher the N fertilizer rate, there will be more chances for leaching out with drainage, especially if the crop does not respond. Delin and Stenberg, [47] checked the role of heavy N fertilization on nitrate leaching and reported that if the N fertilizers were applied at higher rates when crop response was minimal then nitrate was leached down with drainage and nitrate leaching increased exponentially with heavy fertilization. Jia et al. [48] observed that application of higher N fertilizer rates at maize crop showed higher nitrate concentration in leaching water, which makes it dangerous pollution for underground water. Higher nitrogen fertilizer rates, irrigation water, and their interaction affected both leaching and volatilization losses of nitrogen.

Raveendrakumaran et al. [49] evaluated the leaching behavior of various fertilizers on spinach crop and reported that nitrate leaching were inclined type of fertilizers and their application frequency. Frequent split uses of urea are beneficial in reducing nitrate leaching. Klaus et al. [50] observed a negative relation between nitrate leaching risk and legume cover; according to them, the reason was that legume cover has higher land use intensity and decreased plant density by high rates of nitrogen fertilization. Nitrates are most commonly found in agricultural land as a result of inorganic fertilizers i.e. urea.

Nano fertilizers are best options to control nitrate leaching in the agriculture field. Nano fertilizers increase fertilizer use efficiency as it facilitates nutrients availability to the plant leaves. A study by Alimohammadi et al. [51] evaluated effect of urea fertilizer with nano-nitrogen chelate (NNC) on nitrate leaching in soil and it was reported that use of nano nitrogen fertilizer decresed leaching of nitrate to minimize special effects of nitrate leaching in crops, environment and human health as compared with urea. Loss of ammonia gas from urea fertilizer significantly decrease efficiency of urea-nitrogen fertilizer [52]. Latifah et al. [53] reported that when urea is mixed with zeolite and sago wastewater, it is even more beneficial than mixing urea alone because it encourages ammonium and nitrate ions formation over ammonia gas. They confirmed that use of this mixture makes nitrate available within soil and it also improved exchangeable ammonium retention.

2. Nitrogen conversion mechanism

The process of conversion of ammonium (NH_4^+) via nitrite (NO_2^-) into nitrate (NO_3^-) is called nitrification [54]. During nitrification, ammonia is converted into nitrate through oxidation and microorganisms from two different genera are involved in this process. For example, ammonia-oxidizing bacteria (AOB) nitrosomonas,

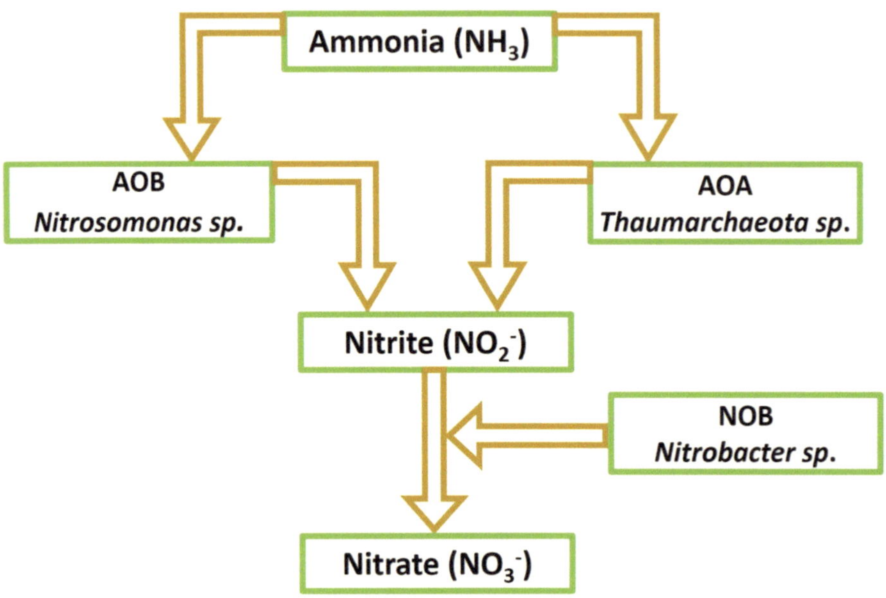

Figure 1.
Nitrogen conversion mechanism.

ammonia-oxidizing archaea (AOA), and nitrite-oxidizing bacteria (NOB) nitrobacter [55]. These bacteria are autotrophic having mutual symbiosis [56].

In the first step ammonium oxidation into nitrite is done by AOB and AOA [57]. The reaction of ammonium oxidation takes place by ammonia monooxygenase (AMO) enzyme [57, 58]. Ammonia-oxidizing archaea (AOA) are usually present in larger numbers in soil but are less important than AOB. In nitrogen-rich agricultural soil systems, AOB are found in abundance, on the other hand, AOA are mostly found in low fertile environments (**Figure 1**) [59, 60].

The second step is the oxidation of nitrite into nitrate (final product). This is done by nitrite-oxidizing bacteria (NOB) (**Figure 1**) i.e., Nitrosomonas. This conversion process of NO_2^- to NO_3^- occurs very speedily hence accumulation of nitrite in soil takes place rarely. Or this whole process of nitrification is done by recently discovered commamox bacteria (complete ammonia-oxidizing bacteria) involves in direct oxidation of ammonia into nitrate [61].

3. Factors affecting nitrate leaching

3.1 Soil texture

Nitrate leaching is strongly inveigled by the texture of soil [62], different soils have different flow rates of nitrate leaching [63]. It is crucial to have knowledge about soil texture while studying nitrate leaching mechanisms. The permeability of water depends upon soil structure and texture [64, 65]. Macro pores within the soil interconnect and make channels for nitrate leaching. Therefore, leaching losses are more in coarse texture or sandy soils just because of their macropores [66]. Clayey and silty textured soils have slow nitrate leaching rates these are favorable for denitrifications. Kasper et al. [67] observed effects of different soil textures on nitrate leaching and reported that rate of nitrate loss was low in silty loam as compared to silty clay loam and sandy loam soil texture. Rainfall is a considerable factor in this leaching while keeping in mind that well-drained soils are prone to high nitrate leaching [68]. Because sandy soils have large pores and high infiltration rate when fertilizers are applied in larger amount, these macropores made the leaching of nitrate easy [69]. So, the texture of soil should be improved for retention of fertilizer and water. This will ultimately reduce nitrate losses, cost of fertilizers, yield loss and protect the environment from pollution.

3.1.1 Season and climate

Season and climate are important factors for nitrate leaching [68]. Nitrate leaching is usually more during time period of late autumn, early spring and winter [70] because in these seasons temperature is low and uptake of nitrate by plant is slow [71]. Due to high precipitations drainage is more in soils. Rainfall during autumn results in the leaching of residuals of fertilizers [72]. During dry summers, nitrate accumulated in surface soil reason is lower uptake of nitrogen and plant growth, this nitrate leached over the upcoming winter. After dry summers, rewetting of dry soils can boost up the rate of mineralization hence more nitrates are leached [32]. Heavy rainfall after fertilizer application can cause leaching in the root zone. On the other hand in tropical areas, summer monsoon rains cause a greater loss of nitrate as compared to winter rains [73].

3.1.2 Agricultural drainage systems

The agricultural drainage systems also conribute in nitrate leaching [74]. Leaching losses are more under efficient drainage systems because of high N mineralization. They also help to shorten the distance traveled by nitrate into rivers and lakes through the soil [75].

3.1.3 Soil organic matter

Soil organic matter and its mineralization have a great role in nitrate leaching [76]. Unfertilized lands or fallow soils have high rate of nitrate leaching as compared to fertilize lands [77].

3.1.4 Earth worms

Earth worms create macrospores when they move along the soil. Those pores help in plant root penetration as well as nitrate is leached quickly through these pores. So nitrate losses are more [78, 79].

4. Strategies to reduce N losses

By adoption of local and scientific means, recovery of nitrogen can be improved that ensure efficient use of fertilizers, irrigation, crop as well as land. It helps to improve beneficial uses of N fertilizer for crops and also reduces losses through leaching [13, 80]. For management of nitrogen following strategies can be used (a)Beneficial use of native soil nitrogen, as well as externally applied N fertilizers [81, 82] (b), conserve nitrogen present in soil by reducing its losses via different mechanisms and to enhance the utilization of nitrogen that is conserved in soils for grown crops (**Table 1**) [97].

4.1 Integrated nitrogen management

Integrated Nitrogen Management (INM) means optimal application of nitrogen inputs i.e. biological nitrogen fixation (BNF), chemical fertilizers, resides of crops, organic fertilizers [98]. Better root growth, higher supply of secondary and micronutrients, and optimum soil environment can be achieved by integrated use of N either Organic or inorganic. Interaction of N with micronutrients and secondary nutrients increases NUE and improves crop yields [99].

4.2 Enhance fertilizer use efficiency

Increasing fertilizer use efficiency (FUE) can be achieved by minimizing nutrient losses. Some fertilizers are based on two important ideas either (a) slow nutrient release or (b) nutrient transformation and reduction in their losses [99, 100].

4.3 Slow-release fertilizer

Slow-release fertilizers have the potential to reduce losses of nitrogen by delaying patterns of N release [101]. They increase NUE and management between

Crop	Sewage Sludge Rate	Effect on Soil Properties	Effect on Crop	Reference
Mung bean	6; 9 and 12 kg m^{-2} (SSA)	decrease pH; increase organic carbon; increase total iron and heavy metals	Increase Contents of Pb, Cd and Ni in grains; increase in plant shoot length, leaf area and plant total biomass	[83]
Cucumber	25; 50; 75; 100 and 124 t ha^{-1}	Increase Soil EC; Increased soluble cations and anions; Increased available phosphorous, micronutrients and heavy metals; decreased soil pH and HCO3-; improved soil physical properties (sandy and calcareous soils)	Increased macro- and microelements in cucumber leaves and fruit; Improved number of cucumber fruits	[84]
Sunflower	9; 18; and 36 Mg dry weight ha^{-1} $yr.^{-1}$		improved vegetative growth of sunflower; enhanced achene yield; decreased achene oil concentration; increased achene N concentration	[85]
Durum Wheat	20; 30; and 40 tons ha^{-1} (dry sludge)		improved grain yield and other components of yields; upsurge in fertility of spike and yield of straw;	[86]
Sunflower	15; 30; 60 and 120 t ha^{-1} (dehydrated SS)	soil pH decreases; soi EC increased; increase mineral N and available phosphorus;	Vegetative growth increased; Increased in dry biomass accumulation; increasedCO_2 Net assimilation; stomatal conductance and decresed transpiration rates;	[87]
Wheat	0;10; 20; 30; 40 and 50 g kg^{-1} (dry sludge)	Increased the soil OM content; decreased pH of soil; increased physicochemical properties of soil	Increased soil fertility and improved crop production; Increased heavy metal accumulation in wheat (except grains Cr and spikes Pb)	[88]
White Radish and Green Beans	22 kg Kala compost with treated waste water	Increased EC and decreased pH; Increased total organic carbon; Increased heavy metal contents except Fe	Increased growth and productivity of vegetables No metal toxicity detected in both crops;	[89]

Crop	Sewage Sludge Rate	Effect on Soil Properties	Effect on Crop	Reference
Wheat	10; 20; 40; 60; 80 and 100 t ha^{-1}	decreased pH; increased soil OM, EC, NPK, Ca, Mg and some other trace metals like Fe, Cu, Mn and Zn	Increased crop yield and yield components as compared to control treatments	[90]
Wheat	control (As 50 mg kg-1); SS (30 g kg^{-1}) + TU (6.5 mM) + As; TU + As; SS + As		SS + TU improve carotenoids, P and K uptake in wheat; SS + TU decreased POD, SOD and APX in wheat under As stress	[91]
wheat-white head cabbage-tomato	10; 20; 30; 40 and 50 tons sludge ha^{-1}		crop yield and yield components were increased (wheat, white head cabbage and tomato) with increasing SS application rates as compared to untreated control	[92]
Wheat	10 ton per hectare		Increase spikelets/ spike and number of grains/spike	[93]
Palak (*Beta vulgaris*)	22.4 g kg^{-1} soil of sewage sludge from 3 locations	significant build-up in DTPA-extractable micronutrient and heavy metal over control; increase of N, P and SOM for production of crop	Increase in micronutrients and heavy metal accumulation; biomass yield of Palak was significantly increased	[94]
Ridge Gourd (*Luffa acutangula* (L.) Roxb.)	Sewage Sludge and Plant Growth-Promoting Rhizobia		improve seedling, plant biochemical response, ridge gourd yield; improved production of crop and recycling of nutrients;	[95]
Barley	0, 4, 8 and 12 t da^{-1} of Sewage Sludge		sewage sludge is effeicinet way to increase yield and metal content (Na, K, and Ca) of grain of barley	[96]

Table 1.
Effect of sewage sludge on soil characteristics, growth and yield of crop.

nitrogen demand of crops and nitrogen supply capacity of soils. Nitrogen losses depend on the form of applied N fertilizers which affects nitrogen available with its recovery [32]. Now a days, neem coated urea fertilizer is used as slow-release N fertilizer [102, 103].

4.3.1 Nitrification inhibitors

Conversion of NH4+ into Nitrate can be inhibited by these inhibitors. With the help of soil MOs, reduce the accumulation of NO3- in soils. Dicyandiamide (DCD) is used as inhibitor (useful in nitrification inhibition) [104]. Neem cake is used as efficient nitrification inhibitor [105]. By using such fertilizers, nitrate leaching can be reduced.

4.3.2 Urease inhibitors

Delay hydrolysis of urease is important because higher amount of nitrogen residual in the soil system can cause problems for environment. UIs help to inhibit urea hydrolysis and reduce nitrate losses [106].

4.3.3 Improved methods of N application

Well-known/common practice broadcasting is the reason for large amount of Nitrogen losses. In rice field where puddling conditions are created, mud balls techniques are used to place urea inside, this enhances recovery of N and crop yield [107]. Deep placement is also a well-known method to avoid N losses [108]. Foliar spray of fertilizers is absorbed by plant surface i.e., urea spray not only improves NUE but also reduced chances of N losses through runoff, denitrifications, and volatilization [81, 109].

4.3.4 Crop rotation and catch crops

Losses of nitrogen in the form of leaching can be reduced by growing cover crops [110–112] or catch crops for example cereal rye, ryegrass, oats, clover, oil-seed radishes etc. Leguminous crops with lower N demand and those having low water requirements like tees are the best options to grow for reducing N-use and nitrate leaching [70, 113].

4.3.5 Treatment of well waters

Wells water should be properly checked either having high nitrate contents or not. If yes then treatments should be done to remove leached nitrate or water should not be used for drinking purposes. It can be used as irrigation water for crops but use of N-fertilizer should be restricted while using well water with higher nitrate contents [114, 115].

5. Sewage sludge: a source of nutrient

Sewage sludge is a residual and semi-solid substance in nature [116], which is a by-product of different treatment stages of municipal (domestic, household) and industrial wastewater effluents, it is also known as bio-solids [116–118]. Untreated domestic wastewater is considered as pollution, and its production is increasing rapidly by urban industrial units which is increasing day by day [119]. The most common way to clean out this waste is its disposal in landfills which is harmful to both environment and public health [120]. Dealing it wisely as a valuable resource can reduce its harmful impacts on the environment. Only 1% of wastewater is found in sewage sludge when it is entered into the sewage treatment plant, digested anaerobically, and wastewater is removed from sludge [121] and if it is treated properly, it becomes bio-solid, organic material that is rich with nutrients [117, 122]. Treatment of sewage sludge is done to make sure that the residual is not the source of pathogens, then recycled sewage sludge is utilized as fertilizer for agricultural crops instead of disposing them off into landfills or water bodies. Sewage sludge has different organic and inorganic elements but a small number of heavy metals are also present in it [123, 124]. Through indiscriminate use, it continues to pollute land and groundwater and enters the food chain via crops and vegetables, damaging the environment.

When sewage sludge is added to the soil, it show beneficial impacts on fertility of soil [125]; it enriches the soil with nitrogen (N), phosphorus (P) and other vital micronutrients [126], improves soil physical, microbial, chemical, and enzymatic properties [127] that ultimately increase crop yield [128]. In different parts of the world, management strategies for sewage sludge are not implemented. Mohamed et al. [87] used dehydrated activated sewage sludge application on sunflower crop. It affected soil properties to increase mineral nitrogen, available phosphorus and soluble potssium. High levels of N and P in sewage sludge explain the beneficial effects on growth of sunflower crop. When Sunflower crop was treated with sludge it showed the highest yield of sunflower seeds. The higher yield was associated with increase in number as well as weight of sunflower seeds.

Throughout the past few years, the use of sewage sludge in agricultural and cropping activities has become broadly accepted as an efficient way to improve crop growth. Khaliq et al., [89] conducted an experiment and it said that 'Kala compost' is a good soil media for production of higher yield in contrast to synthetic fertilizer i.e. NPK. They used "green beans and white radish" as a test crop and the result said that crop yield, total organic carbon, and chlorophyll contents in plants of green beans and white radish were increased when experimental soil recieved "Kala compost" as compared to fertilizer NPK. Soil and crops chemical analysis did not show accumulation of of heavy metals. Baawain et al. [129] and Onwudiwe et al. [130] conducted similar experiments in which their results showed that sewage sludge can improve soil physicochemical properties and it affect crop growth and yields. There are many micro and macronutrients in these fertilizers, which plants uptake to improve their physical and physiological performance and increase soil fertility [131].

The sewage sludge from wastewater treatment systems has become a crucial source of nutrients for urban and peri-urban farmers, predominantly because it is more readily available and cheaper than inorganic fertilizers. Gwenzi et al. [132] experimented to compare the three organic amendments; sewage sludge (SS), sludge biochar (SB), and their combination (SS + SB) to untreatmed control and synthetic fertilizers on maize crop. The results revealed when sewage sludge and sewage sludge biochar is applied to the soil, it significantly improved soil properties, maize growth, plant biomass productivity, and essential nutrients uptake, and reduced uptake of heavy metal. Biochar alone has low nitrogen because nitrogen is volatilized during its production process, so the application of biochar with sludge improves its nutrient content which after application improves soil properties and provides essential nutrients to plants [133].

Compost made from organic municipal solid waste (OMSW) or sewage sludge (SS) provides an opportunity to recycle plant essential nutrients, thus reducing requirement of inorganic/synthetic fertilizers in agriculture. Jamil et al. [90] said that organic fertilizer application (OMSW and SS) is useful in agricultural sustainability. They reported that OMSW and SS consists of several essential nutrients for plants like N, P, K, Zn, Cu, Fe, Mn, and some other trace elements. Debiase et al. [134] confirmed that SS application gives rise to yield of wheat crop 12% more than municipal solid waste. According to him, sewage sludge minimizes nitrogen leaching risk from the soil. Tyagi and Lo, [135] studied that water-activated sludge (WAS) is source of nutrients like nitrogen and phosphorus, when WAS is further solubilized and disintegrated it is converted into ammonia and phosphate possibly resulted in the formation of magnesium ammonium phosphate or struvite; a cost-effective fertilizer.

Sewage sludge amendment (SSA) improved different properties of soil and provide nutrients to crop. Zuo et al. [136] conducted an experiment to effect effects of SSA on soil physical properties, soil chemical properties, sweet sorghum crop yield, and quality in mudflat saline-alkaline soil that was newly reclaimed. Results said that SSA significantly improved plant biomass and crop gross energy content. SSA also elevated soil physical and chemical properties as it decreased soil bulk density, EC, pH, and increased organic carbon contents, water-stable aggregate fraction, CEC, N, and P contents in mudflat soil. Soua and Figueiredo [137] conducted a greenhouse experiment to observe possible effects of different SS biochar doses on soil fertility components and radish crop agronomic development, results said that N concentration in plant leaves was significantly increased that was improved because of higher nitrogen levels in biochar SS. This ultimately increased soil nitrogen in following forms i.e. NO_3^- and NH_4^+. The growth of radish was increased because of nutrients enrichment in the soil amended with biochar, foremost available-P, total-N, and exchangeable cations. According to Chu et al. [138] absorption of nitrogen is more in ammonium form than nitrate as the pH of the soil was almost neutral, neutral pH favors ammonium absorption.

Sewage sludge is used as an organic fertilizer for crop production. A better way to utilize waste is to use it as a resource. Addition of treated sewage sludge into agricultural soils increase fertility; it is a source of nitrogen in soil. Its utilization as an alternative to expensive inorganic fertilizer. But the condition is that SS should be used after its proper treatments, indiscriminate use can cause detrimental effects on soil, groundwater, and food chain.

6. Conclusions

Nitrate leaching is a major problem in certain soils owing to inequitable use of synthetic ferilizers. Moderate application of different organic amendments for example sewage sludge can supply available nitrogen and reduce the use of conventional synthetic fertilizers that may lead to reduced nitrate leaching. Sewage sludge is an organic source of nutrients, that not only provide nutrition to crops but also improve physico-chemical properties of soils and agro-morphological attributes and yields of different crops. It can replace mineral fertilizer in modern cropping systems. Prior application, there should be proper screening of sewage sludge for heavy metals or other toxic elements. Moreover, attention should be paid on nutrients amount as well as their ratio present in the sewage sludge that it can fulfill plant needs for growth and development. In addition, it is recommended for prolonged vegetation period because of slow release of esseantial nutrients. There is also a need for further studies on phytotoxic effects of sewage sludge.

Funding

These is no funding available for this study.

Conflict of interest

"The authors declare no conflict of interest."

Author details

Sidra Sohail[1†], Muhammad Fraz Ali[2,3*†], Usman Zulfiqar[4], Saddam Hussain[2] and Shaharyar Khosa[5]

1 Institute of Soil and Environmental Sciences, University of Agriculture, Faisalabad, Pakistan

2 Department of Agronomy, University of Agriculture, Faisalabad, Pakistan

3 College of Agronomy, Northwest A & F University, Yangling, Shaanxi, China

4 Department of Agronomy, Faculty of Agriculture and Environment, The Islamia University of Bahawalpur, Bahawalpur, Pakistan

5 Faculty of Agricultural Sciences and Technology, Department of Soil Science, Bahauddin Zakariya University, Multan, Punjab, Pakistan

*Address all correspondence to: frazali15@gmail.com

† These authors contributed equally to this chapter.

References

[1] Mahmud K, Panday D, Mergoum A, Missaoui A. Nitrogen losses and potential mitigation strategies for a sustainable agroecosystem. Sustainability. 2021;**13**(4):2400

[2] Duru M, Therond O, Martin G, Martin-Clouaire R, Magne M-A, Justes E, et al. How to implement biodiversity-based agriculture to enhance ecosystem services: A review. Agronomy for Sustainable Development. 2015;**35**(4):1259-1281

[3] Westhoek H, Lesschen JP, Leip A, Rood T, Wagner S, De Marco A, et al. Nitrogen on the table: The influence of food choices on nitrogen emissions and the European environment. NERC/Centre for Ecology & Hydrology. 2015:1-70

[4] Ichihashi Y, Date Y, Shino A, Shimizu T, Shibata A, Kumaishi K, et al. Multi-omics analysis on an agroecosystem reveals the significant role of organic nitrogen to increase agricultural crop yield. Proceedings of the National Academy of Sciences. 2020;**117**(25):14552-14560

[5] Shi J, Yi K, Liu Y, Xie L, Zhou Z, Chen Y, et al. Phospho enol pyruvate carboxylase in Arabidopsis leaves plays a crucial role in carbon and nitrogen metabolism. Plant Physiology. 2015;**167**(3):671-681

[6] Lawlor DW. Plant responses to climate change: Impacts and adaptation. In: Plant Responses to Air Pollution and Global Change. Tokyo, Japan: Springer; 2005. p. 81-88

[7] Lawlor DW, Lemaire G, Gastal F. Nitrogen, plant growth and crop yield. In: Plant Nitrogen. Berlin, Heidelberg: Springer; 2001. p. 343-367

[8] Anas M, Liao F, Verma KK, Sarwar MA, Mahmood A, Chen Z-L, et al. Fate of nitrogen in agriculture and environment: Agronomic, eco-physiological and molecular approaches to improve nitrogen use efficiency. Biological Research. 2020;**53**(1):1-20

[9] Xu G, Fan X, Miller AJ. Plant nitrogen assimilation and use efficiency. Annual Review of Plant Biology. 2012;**63**:153-182

[10] Whalen JK, Kernecker ML, Thomas BW, Sachdeva V, Ngosong C. Soil food web controls on nitrogen mineralization are influenced by agricultural practices in humid temperate climates. CAB Review. 2013;**8**(23):1-18

[11] Yang L, Wang Y, Kobayashi K, Zhu J, Huang J, Yang H, et al. Seasonal changes in the effects of free-air CO2 enrichment (FACE) on growth, morphology and physiology of rice root at three levels of nitrogen fertilization. Global Change Biology. 2008;**14**(8):1844-1853

[12] Fageria NK, Moreira A. The role of mineral nutrition on root growth of crop plants. Advances in Agronomy. 2011;**110**:251-331

[13] Fageria NK, Baligar VC. Enhancing nitrogen use efficiency in crop plants. Advances in Agronomy. 2005;**88**:97-185

[14] Minikaev D, Zurgel U, Tripler E, Gelfand I. Effect of increasing nitrogen fertilization on soil nitrous oxide emissions and nitrate leaching in a young date palm (Phoenix dactylifera L., cv. Medjool) orchard. Agriculture, Ecosystems and Environment. 2021;**319**:107569

[15] Cassman KG, Dobermann A. Nitrogen and the future of agriculture: 20 years on. Ambio. 2022;**51**(1):17-24

[16] Sadaf J, Shah GA, Shahzad K, Ali N, Shahid M, Ali S, et al. Improvements in wheat productivity and soil quality can accomplish by co-application of biochars and chemical fertilizers. Science of the Total Environment. 2017;**607-608**:715-724. [Internet]. DOI: 10.1016/j.scitotenv.2017.06.178

[17] Ribaudo M, Hansen L, Livingston MJ, Mosheim R, Williamson J, Delgado J. Nitrogen in agricultural systems: Implications for conservation policy. SSRN Electronic Journal. 2011;(127):1-82

[18] Ali I, He L, Ullah S, Quan Z, Wei S, Iqbal A, et al. Biochar addition coupled with nitrogen fertilization impacts on soil quality, crop productivity, and nitrogen uptake under double-cropping system. Food and Energy Security. 2020;**9**(3):e208

[19] Vitousek PM, Hättenschwiler S, Olander L, Allison S. Nitrogen and nature. AMBIO: A Journal of the Human Environment. 2002;**31**(2):97-101

[20] Robertson GP, Vitousek PM. Nitrogen in agriculture: Balancing the cost of an essential resource. Annual Review of Environment and Resources. 2009;**34**(1):97-125

[21] Smil V. Nitrogen in crop production: An account of global flows. Global Biogeochemical Cycles. 1999;**13**(2):647-662

[22] Galloway JN, Cowling EB. Reactive nitrogen and the world: 200 years of change. AMBIO: A Journal of the Human Environment. 2002;**31**(2):64-71

[23] Herrera JM, Rubio G, Häner LL, Delgado JA, Lucho-Constantino CA, Islas-Valdez S, et al. Emerging and established technologies to increase nitrogen use efficiency of cereals. Agronomy. 2016;**6**(2):25

[24] Norton J, Ouyang Y. Controls and adaptive management of nitrification in agricultural soils. Frontiers in Microbiology. 2019;**10**:1931

[25] Wakida FT, Lerner DN. Non-agricultural sources of groundwater nitrate: A review and case study. Water Research. 2005;**39**(1):3-16

[26] Zhang Y, Liu J, Mu Y, Pei S, Lun X, Chai F. Emissions of nitrous oxide, nitrogen oxides and ammonia from a maize field in the North China plain. Atmospheric Environment. 2011;**45**(17):2956-2961

[27] Fowler D, Coyle M, Skiba U, Sutton MA, Cape JN, Reis S, et al. The global nitrogen cycle in the twenty-first century. Philosophical Transactions of the Royal Society B: Biological Sciences. 2013;**368**(1621):20130164

[28] Gillette K, Malone RW, Kaspar TC, Ma L, Parkin TB, Jaynes DB, et al. N loss to drain flow and N2O emissions from a corn-soybean rotation with winter rye. Science of the Total Environment. 2018;**618**:982-997

[29] Padilla FM, Gallardo M, Manzano-Agugliaro F. Global trends in nitrate leaching research in the 1960-2017 period. Science of the Total Environment. 2018;**643**:400-413

[30] Azad N, Behmanesh J, Rezaverdinejad V, Abbasi F, Navabian M. Evaluation of fertigation management impacts of surface drip irrigation on reducing nitrate leaching using numerical modeling. Environmental Science and Pollution Research. 2019;**26**(36):36499-36514

[31] Zhang Y, Dong X, Yang X, Munyampirwa T, Shen Y. The lagging movement of soil nitrate in comparison to that of soil water in the 500-cm soil profile. Agriculture, Ecosystems and Environment. 2022;**326**:107811

[32] Cameron KC, Di HJ, Moir JL. Nitrogen losses from the soil/plant system: A review. The Annals of Applied Biology. 2013;**162**(2):145-173

[33] Silva RG, Cameron KC, Di HJ, Smith NP, Buchan GD. Effect of macropore flow on the transport of surface-applied cow urine through a soil profile. Soil Res. 2000;**38**(1):13-24

[34] Lazicki P, Geisseler D. Soil nitrate testing supports nitrogen management in irrigated annual crops. California Agriculture. 2016;**71**(2):90-95

[35] Corrêa RS, White RE, Weatherley AJ. Effect of compost treatment of sewage sludge on nitrogen behavior in two soils. Waste Management. 2006;**26**(6):614-619

[36] Sarker JR, Singh BP, Dougherty WJ, Fang Y, Badgery W, Hoyle FC, et al. Impact of agricultural management practices on the nutrient supply potential of soil organic matter under long-term farming systems. Soil and Tillage Research. 2018;**175**:71-81

[37] Moretti SML, Bertoncini EI, Abreu-Junior CH. Composting sewage sludge with green waste from tree pruning. Science in Agriculture. 2015;**72**:432-439

[38] Rütting T, Aronsson H, Delin S. Efficient use of nitrogen in agriculture. Nutrient Cycling in Agroecosystems. 2018;**110**(1):1-5

[39] Khan MS, Naveed M, Qadir MF, Bashir MA, Rafique M, Siddiqui MH, et al. Combined effect of animal manures and Di-ammonium phosphate (DAP) on growth, physiology, root nodulation and yield of chickpea. Agronomy. 2022;**12**(3):674

[40] Kimetu JM, Mugendi DN, Palm CA, Mutuo PK, Gachengo CN, Bationo A, et al. Nitrogen fertilizer equivalencies of organics of differing quality and optimum combination with inorganic nitrogen source in Central Kenya. Nutrient Cycling in Agroecosystems. 2004;**68**(2):127-135

[41] Spiertz JHJ. Nitrogen, sustainable agriculture and food security: A review. Agronomy for Sustainable Development. 2010;**30**(1):43-55

[42] Khan A, Singh J, Upadhayay VK, Singh AV, Shah S. Microbial biofortification: A green technology through plant growth promoting microorganisms. In: Sustainable Green Technologies for Environmental Management. Singapore: Springer; 2019. pp. 255-269

[43] Rahman KM, Zhang D. Effects of fertilizer broadcasting on the excessive use of inorganic fertilizers and environmental sustainability. Sustainability. 2018;**10**(3):759

[44] Hazra G. Different types of eco-friendly fertilizers: An overview. Sustainability Environment. 2016;**1**(1):54

[45] Panhwar QA, Ali A, Naher UA, Memon MY. Fertilizer management strategies for enhancing nutrient use efficiency and sustainable wheat production. In: Organic Farming. Elsevier; 2019. pp. 17-39

[46] Gu L, Liu T, Jun Z, Dong S, LIU P, Zhang J, et al. Nitrate leaching of winter wheat grown in lysimeters as affected by fertilizers and irrigation on the North China plain. Journal of Integrative Agriculture. 2015;**14**(2):374-388

[47] Delin S, Stenberg M. Effect of nitrogen fertilization on nitrate leaching in relation to grain yield response on loamy sand in Sweden. European Journal of Agronomy. 2014;**52**:291-296

[48] Jia X, Shao L, Liu P, Zhao B, Gu L, Dong S, et al. Effect of different nitrogen and irrigation treatments on yield and nitrate leaching of summer maize (Zea mays L.) under lysimeter conditions. Agricultural Water Management. 2014;**137**:92-103

[49] Raveendrakumaran B, Grafton M, Jeyakumar P, Bishop P, Davies C. Comparative Evaluation of Controlled Release Fertilisers for Nitrate Leaching. In: Nutrient Management in Farmed Landscapes. Palmerston North, New Zealand: Fertilizer and Lime Research Centre, Massey University; 2020. p. 1-5

[50] Klaus VH, Kleinebecker T, Hölzel N, Blüthgen N, Boch S, Müller J, et al. Nutrient concentrations and fibre contents of plant community biomass reflect species richness patterns along a broad range of land-use intensities among agricultural grasslands. Perspectives in Plant Ecology, Evolution and Systematics. 2011;**13**(4):287-295

[51] Alimohammadi M, Panahpour E, Naseri A. Assessing the effects of urea and nano-nitrogen chelate fertilizers on sugarcane yield and dynamic of nitrate in soil. Soil Science & Plant Nutrition. 2020;**66**(2):352-359

[52] Wallace AJ, Armstrong RD, Grace PR, Scheer C, Partington DL. Nitrogen use efficiency of 15N urea applied to wheat based on fertiliser timing and use of inhibitors. Nutrient Cycling in Agroecosystems. 2020;**116**(1):41-56

[53] Latifah O, Ahmed OH, Muhamad AMN. Reducing ammonia loss from urea and improving soil exchangeable ammonium and available nitrate in non waterlogged soils through mixing zeolite and sago (Metroxylon sagu) waste water. International Journal of Physics and Science. 2011;**6**(4):866-870

[54] Ghaly AE, Ramakrishnan VV. Nitrification of urea and assimilation of nitrate in saturated soils under aerobic conditions. American Journal of Agricultural and Biological Sciences. 2013;**8**(4):330-342

[55] Sedlacek CJ, McGowan B, Suwa Y, Sayavedra-Soto L, Laanbroek HJ, Stein LY, et al. A physiological and genomic comparison of Nitrosomonas cluster 6a and 7 ammonia-oxidizing bacteria. Microbial Ecology. 2019;**78**(4):985-994

[56] Stein LY, Klotz MG. The nitrogen cycle. Current Biology. 2016;**26**(3):R94-R98

[57] Ding X, Lan W, Li Y, Yan A, Katayama Y, Koba K, et al. An internal recycling mechanism between ammonia/ammonium and nitrate driven by ammonia-oxidizing archaea and bacteria (AOA, AOB, and Comammox) and DNRA on Angkor sandstone monuments. International Biodeterioration & Biodegradation. 2021;**165**:105328

[58] Huang X, Xu Y, He T, Jia H, Feng M, Xiang S, et al. Ammonium transformed into nitrous oxide via nitric oxide by pseudomonas putida Y-9 under aerobic conditions without hydroxylamine as intermediate. Bioresource Technology. 2019;**277**:87-93

[59] Di HJ, Cameron KC, Shen J-P, Winefield CS, O'Callaghan M, Bowatte S, et al. Nitrification driven by bacteria and not archaea in nitrogen-rich grassland soils. Nature Geoscience. 2009;**2**(9):621-624

[60] Di HJ, Cameron KC, Sherlock RR, Shen J-P, He J-Z, Winefield CS. Nitrous oxide emissions from grazed grassland as affected by a nitrification inhibitor, dicyandiamide, and relationships with ammonia-oxidizing bacteria and archaea. Journal of Soils and Sediments. 2010;**10**(5):943-954 [Internet] Available from: http://link.springer.com/10.1007/s11368-009-0174-x

[61] Daims H, Lebedeva EV, Pjevac P, Han P, Herbold C, Albertsen M, et al. Complete nitrification by Nitrospira bacteria. Nature. 2015;**528**(7583):504-509

[62] Haider G, Steffens D, Moser G, Müller C, Kammann CI. Biochar reduced nitrate leaching and improved soil moisture content without yield improvements in a four-year field study. Agriculture Ecosystem. 2017;**237**:80-94. [Internet]. DOI: 10.1016/j.agee.2016.12.019

[63] Zhu B, Wang T, Kuang F, Luo Z, Tang J, Xu T. Measurements of nitrate leaching from a hillslope cropland in the Central Sichuan Basin, China. Soil Science Society of America Journal. 2009;**73**(4):1419-1426

[64] Murray RS, Grant CD. The impact of irrigation on soil structure. Land and Water Australia. 2007:1-31

[65] Guimarães RML, Ball BC, Tormena CA, Giarola NFB, da Silva ÁP. Relating visual evaluation of soil structure to other physical properties in soils of contrasting texture and management. Soil and Tillage Research. 2013;**127**:92-99

[66] Gerhard L, Puhlmann H, Vogt M, Luster J. Phosphorus leaching from naturally structured Forest soils is more affected by soil properties than by drying and rewetting. Frontiers in Forests and Global Change. 2021;**4**:49

[67] Kasper M, Foldal C, Kitzler B, Haas E, Strauss P, Eder A, et al. N2O emissions and NO3– leaching from two contrasting regions in Austria and influence of soil, crops and climate: A modelling approach. Nutrient cycling in agroecosystems. 2019;**113**(1):95-111 [Internet] Available from: http://link.springer.com/10.1007/s10705-018-9965-z

[68] Osman KT. Management of Soil Problems. Cham: Springer International Publishing; 2018. Available from: http://link.springer.com/10.1007/978-3-319-75527-4

[69] Decock C, Garland G, Suddick EC, Six J. Season and location–specific nitrous oxide emissions in an almond orchard in California. Nutrient Cycling in Agroecosystems. 2017;**107**(2):139-155

[70] Francis GS. Management practices for minimising nitrate leaching after ploughing temporary leguminous pastures in Canterbury, New Zealand. Journal of Contaminant Hydrology. 1995;**20**(3-4):313-327

[71] Jabloun M, Schelde K, Tao F, Olesen JE. Effect of temperature and precipitation on nitrate leaching from organic cereal cropping systems in Denmark. European Journal of Agronomy. 2015;**62**:55-64

[72] Pedersen BN, Eriksen J, Christensen BT, Sørensen P. Fertilizer replacement value and leaching of nitrogen applied to spring barley in cattle deep litter: A 3-year lysimeter study. Soil and Tillage Research. 2021;**209**:104954

[73] Mastrocicco M, Colombani N, Soana E, Vincenzi F, Castaldelli G. Intense rainfalls trigger nitrite leaching in agricultural soils depleted in organic matter. Science Total Environment. 2019;**665**:80-90

[74] Kulhavý Z, Doležal F, Fučík P, Kulhavý F, Kvítek T, Muzikář R, et al. Management of agricultural drainage systems in the Czech Republic. Irrigation and Drainage: The journal of the International Commission on Irrigation and Drainage. 2007;**56**(S1):S141-S149

[75] Castellano MJ, Archontoulis SV, Helmers MJ, Poffenbarger HJ, Six J. Sustainable intensification of agricultural drainage. Nat Sustainability. 2019;**2**(10):914-921

[76] Ashraf MN, Hu C, Wu L, Duan Y, Zhang W, Aziz T, et al. Soil and microbial biomass stoichiometry regulate soil organic carbon and nitrogen mineralization in rice-wheat rotation subjected to long-term fertilization. Journal of Soils and Sediments. 2020;**20**(8):3103-3113

[77] Sharma GK, Khan SA, Shrivastava M, Bhattacharyya R, Sharma A, Gupta N, et al. Phycoremediated N-fertilization approaches on reducing environmental impacts of agricultural nitrate leaching. Journal of Cleaner Production. 2022;**345**:131120

[78] Shourie A, Singh A. Impact of climate change on soil fertility. In: Climate Change and the Microbiome. Cham: Springer International Publishing; 2021. pp. 49-62

[79] Almusaed A. Introduction on Growing Media (Soil). In: Biophilic and Bioclimatic Architecture. London: Springer; 2011. p. 85-94

[80] Leghari SJ, Wahocho NA, Laghari GM, HafeezLaghari A, Mustafa Bhabhan G, Hussain Talpur K, et al. Role of nitrogen for plant growth and development: A review. Advances in Environmental Biology. 2016;**10**(9):209-219

[81] Yadav MR, Kumar R, Parihar CM, Yadav RK, Jat SL, Ram H, et al. Strategies for improving nitrogen use efficiency: A review. Agricultural Reviews. 2017;**38**(1):29-40

[82] Singh A, Kumar A, Jaswal A, Singh M, Gaikwad DS. Nutrient use efficiency concept and interventions for improving nitrogen use efficiency. Plant Architecture. 2018;**18**(1):1015-1023

[83] Singh RP, Agrawal M. Effect of different sewage sludge applications on growth and yield of Vigna radiata L. field crop: Metal uptake by plant. Ecological Engineering. 2010;**36**(7):969-972 [Internet] Available from: https://www.sciencedirect.com/science/article/pii/S0925857410000686

[84] Hussein AHA. Impact of sewage sludge as organic manure on some soil properties, growth, yield and nutrient contents of cucumber crop. Journal of Applied Sciences. 2009;**9**(8):1401-1411

[85] Koutroubas SD, Antoniadis V, Damalas CA, Fotiadis S. Sunflower growth and yield response to sewage sludge application under contrasting water availability conditions. Industrial Crops and Products. 2020;**154**:112670 [Internet]. Available from: https://www.sciencedirect.com/science/article/pii/S0926669020305860

[86] Tamrabet L, Bouzerzour H, Kribaa M, Makhlouf M. The effect of sewage sludge application on durum wheat (Triticum durum). International Journal of Agriculture and Biology. 2009;**11**(6):741-745

[87] Mohamed B, Mounia K, Aziz A, Ahmed H, Rachid B, Lotfi A. Sewage sludge used as organic manure in Moroccan sunflower culture: Effects on certain soil properties, growth and yield components. Science Total Environment. 2018;**627**:681-688

[88] Eid EM, Alrumman SA, El-Bebany AF, Fawy KF, Taher MA, Hesham AE-L, et al. Evaluation of the potential of sewage sludge as a valuable fertilizer for wheat (Triticum aestivum L.) crops. Environmental Science and Pollution Research. 2019;**26**(1):392-401. [Internet]. DOI: 10.1007/s11356-018-3617-3

[89] Khaliq A, Jaffar S, Al-Busaidi A, Ahmed M, Al-Wardy M, Agrama H, et al. The effect of municipal sewage sludge on the quality of soil and crops. International Journal of Recycling of Organic Waste in Agriculture. 2017;**6**(4):289-299

[90] Jamil M, Qasim M, Umar M. Utilization of sewage sludge as organic fertilizer in sustainable agriculture. Journal of Applied Sciences. 2006;**6**(3):531-535

[91] Mansoora N, Kausar S, Amjad SF, Yaseen S, Shahid H, tul Kubra K, et al. Application of sewage sludge combined with thiourea improves the growth and yield attributes of wheat (Triticum aestivum L.) genotypes under arsenic-contaminated soil. PLoS One. 2021;**16**(11):e0259289

[92] Özyazici MA. Effects of sewage sludge on the yield of plants in the rotation system of wheat-white head cabbage-tomato. Eurasian Journal of Soil Science. 2013;**2**(1):35-44

[93] Hammad HM, Khaliq A, Ashfaq A, Aslam M, Malik AH, Farhad W, et al. Influence of different organic manures on wheat productivity. International Journal of Agriculture and Biology. 2011;**13**(1):137-140

[94] Roy T, Singh RD, Biswas DR, Patra AK. Effect of sewage sludge and inorganic fertilizers on productivity and micronutrients accumulation by Palak (Beta vulgaris) and their availability in a Typic Haplustept. Journal. Indian Society of Soil Science. 2013;**61**(3):207-218

[95] Kumar V, Eid EM, Al-Bakre DA, Abdallah SM, Širić I, Andabaka Ž, et al. Combined use of sewage sludge and plant growth-promoting rhizobia improves germination, biochemical response and yield of ridge gourd (Luffa acutangula (L.) Roxb.) under field conditions. Agriculture. 2022;**12**(2):173

[96] Angin I, Yaganoglu AV. Effects of sewage sludge application on yield, yield parameters and heavy metal content of barley grown under arid climatic conditions. International Journal of Agriculture and Biology. 2012;**14**(5):811-815

[97] Manu MK, Li D, Liwen L, Jun Z, Varjani S, Wong JWC. A review on nitrogen dynamics and mitigation strategies of food waste digestate composting. Bioresource Technology. 2021;**334**:125032

[98] Khan MN, Ijaz M, Ali Q, Ul-Allah S, Sattar A, Ahmad S. Biological Nitrogen Fixation in Nutrient Management. In: Agronomic Crops. Singapore: Springer; 2019. p. 127-147

[99] Dimkpa CO, Fugice J, Singh U, Lewis TD. Development of fertilizers for enhanced nitrogen use efficiency–trends and perspectives. Sci Total Environ. 2020;**731**:139113

[100] Xiang YAN, Jin J-Y, Ping HE, Liang M. Recent advances on the technologies to increase fertilizer use efficiency. Agricultural Sciences in China. 2008;7(4):469-479

[101] Naz MY, Sulaiman SA. Slow release coating remedy for nitrogen loss from conventional urea: A review. Journal of Controlled Release. 2016;**225**:109-120

[102] Singh A, Jaswal A, Singh M. Impact of neem coated urea on rice yield and nutrient use efficiency (NUE). Agricultural Reviews. 2019;**40**(1):70-74

[103] Ghafoor I, Habib-ur-Rahman M, Ali M, Afzal M, Ahmed W, Gaiser T, et al. Slow-release nitrogen fertilizers enhance growth, yield, NUE in wheat crop and reduce nitrogen losses under an arid environment. Environmental Science and Pollution Research. 2021;**28**(32):43528-43543

[104] Zhang H, Zhao X, Zhang L, Shen R. Effects of liming and dicyandiamide (DCD) application on soil pH and nitrification of acidic red soil. Acta Pedologica Sinica. 2021;**58**(1):169-179

[105] Singh B. Neem coated urea as a source of nitrogen for plants. Journal of Eco-friendly Agriculture. 2019;**14**(1):43-54

[106] Wang D, Guo L, Zheng L, Zhang Y, Yang R, Li M, et al. Effects of nitrogen fertilizer and water management practices on nitrogen leaching from a typical open field used for vegetable planting in northern China. Agricultural Water Management. 2019;**213**:913-921

[107] Panda D, Nayak AK, Mohanty S. Nitrogen management in rice. Oryza. 2019;**56**:125-135

[108] Maris SC, Abalos D, Capra F, Moscatelli G, Scaglia F, Reyes GEC, et al. Strong potential of slurry application timing and method to reduce N losses in a permanent grassland. Agriculture, Ecosystems and Environment. 2021;**311**:107329

[109] Zhang M, Li X, Wang H, Huang Q. Comprehensive analysis of grazing intensity impacts soil organic carbon : A case study in typical steppe of Inner Mongolia, China. Applied Soil Ecology. 2018;**129**:1-12

[110] Rocha KF, de Souza M, Almeida DS, Chadwick DR, Jones DL, Mooney SJ, et al. Cover crops affect the partial nitrogen balance in a maize-forage cropping system. Geoderma. 2020;**360**:114000

[111] Abdalla M, Hastings A, Cheng K, Yue Q, Chadwick D, Espenberg M, et al. A critical review of the impacts of cover crops on nitrogen leaching, net greenhouse gas balance and crop productivity. Global Change Biology. 2019;**25**(8):2530-2543

[112] Justes E, Rechauchère O, Chemineau P. The use of cover crops to reduce nitrate leaching: Effect on the water and nitrogen balance and other ecosystem services. INRA. 2012:1-68

[113] Carey PL, Cameron KC, Di HJ, Edwards GR. Does sowing an oats catch crop reduce nitrate leaching from urine deposition following simulated winter forage grazing? – a growth chamber experiment. Plant and Soil. 2018;**431**(1-2):37-52

[114] Zhang H, Hu K, Zhang L, Ji Y, Qin W. Exploring optimal catch crops for reducing nitrate leaching in vegetable greenhouse in North China. Agricultural Water Management. 2019;**212**:273-282

[115] Olesen JE, Børgesen CD, Hashemi F, Jabloun M, Bar-Michalczyk D, Wachniew P, et al. Nitrate leaching losses from two Baltic Sea catchments under scenarios of changes in land use, land management and climate. Ambio. 2019;**48**(11):1252-1263

[116] Negi P, Verma H, Singh SP, Mahapatra BS, Jatav HS. Global Scenario of Sewage- Sludge Management. In: Sustainable Management and Utilization of Sewage Sludge. Cham: Springer International Publishing; 2022. p. 383-401

[117] Kumar V, Chopra AK, Kumar A. A review on sewage sludge (biosolids) a resource for sustainable agriculture. Archives of Agriculture and Environmental Science. 2017;**2**(4):340-347

[118] Delibacak S, Voronina L, Morachevskaya E. Use of sewage sludge in agricultural soils: Useful or harmful. Eurasian Journal of Soil Science. 2020;**9**(2):126-139

[119] Yin H, Islam MS, Ju M. Urban river pollution in the densely populated city of Dhaka, Bangladesh: Big picture and rehabilitation experience from other developing countries. Journal of Cleaner Production. 2021;**321**:129040

[120] Alabi OA, Ologbonjaye KI, Awosolu O, Alalade OE. Public and environmental health effects of plastic wastes disposal: A review. Journal of Toxicology Risk Assess. 2019;**5**(021):1-13

[121] Núñez-Delgado A, Pousada-Ferradás Y, Álvarez-Rodríguez E, Fernández-Sanjurjo MJ, Conde-Cid M, Nóvoa-Muñoz JC, et al. Effects of microbiological and non-microbiological treatments of sewage sludge on antibiotics as emerging pollutants present in wastewater: A review. In: Microbial Wastewater Treatment. Elsevier; 2019. p. 1-17

[122] Kominko H, Gorazda K, Wzorek Z. The possibility of organo-mineral fertilizer production from sewage sludge. Waste and Biomass Valorization. 2017;**8**(5):1781-1791

[123] Iticescu C, Georgescu LP, Murariu G, Circiumaru A, Timofti M. The characteristics of sewage sludge used on agricultural lands. AIP Conference Proceedings. 2018:20001

[124] Geng H, Xu Y, Zheng L, Gong H, Dai L, Dai X. An overview of removing heavy metals from sewage sludge: Achievements and perspectives. Environmental Pollution. 2020;**266**:115375

[125] Callegari A, Capodaglio AG. Properties and beneficial uses of (bio) chars, with special attention to products from sewage sludge pyrolysis. Resources. 2018;**7**(1):20

[126] Bourioug M, Alaoui-Sossé L, Laffray X, Raouf N, Benbrahim M, Badot P-M, et al. Evaluation of sewage sludge effects on soil properties, plant growth, mineral nutrition state, and heavy metal distribution in European larch seedlings (Larix decidua). Arabian Journal for Science and Engineering. 2014;**39**(7):5325-5335

[127] Singh V, Diwedi A. Chapter-1 application of sewage sludge influencing soil health and crop production. Current Research In Soil Fertility. 2019;**19**:1

[128] Taylor RP, Jones CLW, Laubscher RK. Agricultural fertiliser from brewery effluent–the recovery of nutrients from the biomass of activated sludge and high rate algal pond treatment systems. Water Supply. 2021;**21**(5):1939-1952

[129] Baawain MS, Al-Omairi A, Choudri BS. Characterization of domestic wastewater treatment in Oman from three different regions and current implications of treated effluents. Environmental Monitoring and Assessment. 2014;**186**(5):2701-2716

[130] Onwudiwe N, Benedict OU, Ogbonna PE, Ejiofor EE. Municipal solid waste and NPK fertilizer effects on soil physical properties and maize performance in Nsukka, Southeast Nigeria. African Journal of Biotechnology. 2014;**13**(1):68-75

[131] Zhao XL, Mu ZJ, Cao CM, Wang DY. Growth and heavy-metal uptake by

lettuce grown in soils applied with sewage sludge compost. Communications in Soil Science and Plant Analysis. 2012;**43**(11):1532-1541

[132] Gwenzi W, Muzava M, Mapanda F, Tauro TP. Comparative short-term effects of sewage sludge and its biochar on soil properties, maize growth and uptake of nutrients on a tropical clay soil in Zimbabwe. Journal of Integrative Agriculture. 2016;**15**(6):1395-1406

[133] Li S, Wang S, Shangguan Z. Combined biochar and nitrogen fertilization at appropriate rates could balance the leaching and availability of soil inorganic nitrogen. Agriculture, Ecosystems and Environment. 2019;**276**:21-30

[134] Debiase G, Montemurro F, Fiore A, Rotolo C, Farrag K, Miccolis A, et al. Organic amendment and minimum tillage in winter wheat grown in Mediterranean conditions: Effects on yield performance, soil fertility and environmental impact. European Journal of Agronomy. 2016;**75**:149-157

[135] Tyagi VK, Lo S-L. Sludge: A waste or renewable source for energy and resources recovery? Renewable and Sustainable Energy Reviews. 2013;**25**:708-728

[136] Zuo W, Gu C, Zhang W, Xu K, Wang Y, Bai Y, et al. Sewage sludge amendment improved soil properties and sweet sorghum yield and quality in a newly reclaimed mudflat land. Science Total Environment. 2019;**654**:541-549 [Internet]. Available from: https://www.sciencedirect.com/science/article/pii/S004896971834484X

[137] Sousa AATC, Figueiredo CC. Sewage sludge biochar: Effects on soil fertility and growth of radish. Biological Agriculture and Horticulture. 2016;**32**(2):127-138. [Internet]. DOI: 10.1080/01448765.2015.1093545

[138] Chu Q, Xue L, Singh BP, Yu S, Müller K, Wang H, et al. Sewage sludge-derived hydrochar that inhibits ammonia volatilization, improves soil nitrogen retention and rice nitrogen utilization. Chemosphere. 2020;**245**:125558 [Internet]. Available from: https://www.sciencedirect.com/science/article/pii/S0045653519327985

Chapter 4

Occurrence, Bioaccumulation and Effects of Legacy and Emerging Brominated Retardants in Earthworms

Jean-Philippe Bedell, Claudia Coelho, Olivier Roques, Anais Venisseau, Philippe Marchand and Yves Perrodin

Abstract

The presence of former brominated flame retardants and "emerging" brominated flame retardants (BFRs and e-BFRs) in soils is well documented, but the presence, metabolism and uptake of them in earthworm species are much less. Polybrominated diphenyl ethers (PBDEs) and hexabromocyclododecanes (HBCDDs) are the most abundant "legacy" BFRs in soils. Earthworms are a good bioindicator, presenting an integrated view of soil chemical pollution. They bioaccumulate BFRs passively by dermal absorption, and actively through soil ingestion. However, such information is only available for a limited number of species, mostly for *Eisenia fetida,* which shows high bioaccumulation factors (>2). Most of the ecotoxicity studies on earthworms have been done using PBDEs or HBCDDs. PBDEs were reported to effect changes in enzyme activities, which induced oxidative stress and caused metabolic perturbations in some earthworm species. In *E. fetida,* contaminant bioaccumulation is influenced by the lipid and protein contents of tissues, but several different processes (uptake, depuration, metabolism and isomerization) also contribute to the observed tissue levels. To evaluate and manage the risks posed by these chemicals to terrestrial ecosystems, it is important to better understand the transfer processes of emerging brominated flame retardants in earthworms, as well as the potential trophic biomagnification.

Keywords: bioaccumulation, brominated flame retardants, earthworms, *Eisenia fetida*, emerging brominated flame retardants, PBDEs

1. Introduction

Brominated flame retardants (BFRs) are in demand on the market because of their relatively low cost and efficient fire-resistance performance [1], notably because the BFRs' incorporation generates halogen atoms of the parent compound that can delay or "retard" the development of a fire [2]. Thus, BFRs are used in different product categories, such as textiles, plastics, building materials and very often in electronic equipment [3–5].

Some BFRs have been classified as "traditional" or "legacy" BFRs in contrast to the "novel" or "emerging" BFRs (e-BFRs) [6, 7]. Legacy BFRs mainly refer to the chemicals that are no longer used or commercialized, such as polybrominated biphenyls (PBBs), polybrominated diphenyl ethers (PBDEs), hexabromocyclododecanes (HBCDDs) and tetrabromobisphenol A (TBBPA). Among the e-BFRs, we can list decabromodiphenyl ethane (DBDPE), bis(2,4,6-tribromophenoxy) ethane (BTBPE), bis(2-ethylhexyl) tetrabromophthalate (BEH-TEBP), 2-ethylhexyl-2,3,4,5-tetrabromobenzoate (EH-TBB), hexabromobenzene (HBBz), pentabromobenzene (PBBz), pentabromotoluene (PBT) and pentabromoethylbenzene (PBEB). These e-BFRs remain poorly studied and documented [8], but they present similar physicochemical properties as other persistent organic pollutants (POPs), notably polychlorobiphenyls (PCBs), and may therefore pose a potentially hazardous risk to ecosystems. Thus, BFRs are characterized by medium to low vapor pressures as well as low water solubility and high volatility, with high (> 6) octanol–water partition coefficients (Kow) [5, 6]. Such properties indicate that they are lipophilic. The physicochemical and biochemical properties of BFRs combined with their resistance to chemical and biological degradation make these molecules persistent and easily incorporated in every compartment of the ecosystem [9, 10]. In Ref. [11], it underlined their persistence and bioaccumulative characteristics as well as their potential toxicity to living organisms. In addition, exposure to BFRs results in negative health effects for animals and humans [9, 10, 12]. PBBs, PBDEs and HBCDDs were the most widely produced and used BFRs. Due to volatilization, weathering and adsorption processes, notably on dusts, BFRs can migrate to environmental compartments and can be released into the air [13]. In addition, at the end of a product life, the incorporated BFRs may be released during waste disposal. In Ref. [14], they classified ΣPBDEs by land use in many different soils, with values ranging between 0.05 and 800 $ng.g^{-1}$ for urban soils and between 3 and 40,000 $ng.g^{-1}$ for electronic and electrical waste contaminated soils. The variability observed between several measurements demonstrates the ubiquitous dispersion of these compounds across the world and the great difficulty in comparing different soil samples. This difficulty is linked to the types and characteristics of soils (such as organic matter (OM) or clay content), which may be factors influencing the retention/adsorption/fixing of the BFRs, as well as the different potential BFR sources (industrial, waste disposal and urban) and their atmospheric transport and dispersion. In the literature, if PBDEs are well documented, there is a lack of studies and data regarding the environmental fate of PBBs and HBCDDs, notably in soils. Despite its extensive use, information on TBBPA concentrations in soils [15, 16] is also scarce compared to data available for PBDEs, for example, see in [14–18]. Legacy BFRs are still present in the environment and are obviously well dispersed in a lot of different soils.

Electronic waste (E-waste) dismantling areas seem to be a major source of e-BFRs in the environment, which are dispersed by atmospheric transport followed by deposition on land surfaces. The hydrophobic nature of e-BFRs is confirmed by the observed log K_{ow} values that are greater than 3, which is similar to the legacy BFRs. These compounds are therefore highly lipophilic, which can explain their potential to bioaccumulate in organisms. This bioaccumulation has already been reported in the literature, mainly for aquatic animals, but similar data on e-BFRs in soils, earthworms and plants are scarce. Of the e-BFRs, DBDPE was the dominant compound detected in the Arctic soil (average; 9.1 $pg.g^{-1}$ dw; [19]) or in the rhizosphere or non-rhizosphere of an e-waste soil in South China [20]. Other studies also reported that the DBPDE content was higher than BDE 209 in the Chinese background forest soil [21]. In addition, HBBz also generally occurred at high concentrations in soils, ranging from close to, or above 1 ng.g^{-1} dw to a maximum of ~10 ng.g^{-1} dw depending on the type of soil [14, 17, 21–24]. The other compounds, such as PPBz, PBT and PBEB, showed slightly lower values on average, except at some contaminated sites which are generally where most measurements have been made (e-waste or disposal sites) [23, 25, 26]. The physicochemical characteristics, patterns of environmental contamination and toxicity effects of e-BFRs are close to those of PBDEs. These analogies suggest that availability and bioaccumulation processes might be similar to those of PBDEs, as well as to those of PCBs or polychlorinated dibenzo-p-dioxins/polychlorinated dibenzofurans (PCDD/Fs).

The present work focuses on brominated flame retardants (BFRs) in particular, because certain "new or emerging" BFRs (e-BFRs) have replaced a number of legacy brominated and chlorinated POPs. Moreover, two interesting reviews of BFRs in soil [17] and in wildlife [27] highlighted the requirement for more research, notably on e-BFRs, in order to obtain a better understanding of their environmental fate and dispersion, notably in different environmental compartments. In contrast to many invertebrate organisms that are only indirectly exposed to contaminants through the food chain, invertebrates in the soil are good "sentinel organisms" for the chemical contamination of soil [28] because they are in direct contact with soils, soil pore water and their food sources which generally consist of organic matter, algae, fungi and small microorganisms including bacteria. Thus, one of them, earthworms represent the largest biomass fraction of soils and tend to migrate over very short distances, illustrating again the fact that they are a good candidate for integrating soil chemical pollution [29]. Thus, this work gives concentrations and elements on bioaccumulation of legacy and also with many values on e-BFRs in earthworms which are less developed in bibliography. Moreover, we also address an aspect that is less emphasized in the review articles: toxicological effects of these compounds on some species of earthworms, as well as the "metabolic effect" (physiological or biochemical mechanisms) of these pollutants on earthworms.

2. Potential bioaccumulation and effects of BFRs on earthworms

When soils are contaminated by BFRs, they automatically become a source for the transfer of these pollutants to biotic compartments, and notably to an important key organism of the soil ecosystem: earthworms.

2.1 Earthworms as a bioindicator

Earthworms are also called "ecosystem engineers" due to their role: (i) in soil formation, (ii) in organic matter breakdown processes, (iii) in soil aeration as well and (iv) in nutrient cycling [29]. Consequently, earthworms are good indicators of land-use impact and soil fertility, and affect the physical, chemical and microbiological properties of the soil by their roles in the pedogenesis of the soil profile [30]. In addition, earthworms also increase the mineralization and humification of soil organic matter (OM) by consumption, respiration and passage through their gut [31]. Their castings stimulate an increase in OM breakdown activity and microbial mass, while also improving the mobilization of nutrients [32]. Furthermore, their burrowing activities significantly promote the increase in water infiltration capacity and aeration of soils [33]. Unfortunately, these activities also result in the direct exposure of earthworms to both organic and inorganic contaminants that are adsorbed in the soil OM particles that constitute a major feed source for these terrestrial invertebrates [34–36].

Due to their interactions with soil, earthworms can be considered as valuable bioindicators, as they are very sensitive to the presence of contaminants in the soil and are significantly affected by them [37]. Earthworms are exposed to soil contaminants through different routes of exposure. They have direct contact with the soil pore water and through that, with the dissolved contaminants present in this water. Earthworm skin is renowned for being very permeable to water [38] and this contact with the skin represents a main route for contaminant uptake [39, 40]. Moreover, soil ingested by earthworms results in continuous exposure to contaminants adsorbed in the solid particles passing through their digestive tract [41].

A species of earthworm, the *Eisenia fetida* (*E. fetida*), is probably the most used standard test organism in terrestrial ecotoxicology evaluations, due to its rapid life cycle, and easily controlled breeding and reproduction [42]. Most of the works on BFR bioaccumulation, effects or metabolism evaluations have been carried out using *E. fetida*, with a small amount being carried out on other earthworms (**Tables 1** and **2**). Most of these studies have involved legacy BFRs, notably PBDEs (**Table 1**), and few results are available for HBCDDs or e-BFRs (**Tables 2** and **3**).

2.2 Bioaccumulation of BFRs in earthworms

Earthworms have the capacity to accumulate organic (and inorganic) contaminants present in soils [57]. Specifically, studies have demonstrated that several earthworm species are able to accumulate POPs, such as PCBs, BFRs, pharmaceuticals, detergent metabolites, polycyclic aromatic hydrocarbons (PAHs) and pesticides [58–60]. Earthworms can accumulate POPs passively by dermal absorption, and actively through soil ingestion. If the relative importance of dermal and dietary exposure depends on the individual contaminant, some results have highlighted that the importance of the dietary pathway for accumulation increases as a function of the hydrophobicity of the contaminant [39, 61]. In Ref. [40], they evaluated the importance of different pathways for metal uptake in *Lumbricus rubellus*. They concluded that the main route to metal accumulation was dermal absorption because ingestion via pore water uptake represented only a small contribution [40]. In a study with four different species (*Eisenia andrei*, *Eisenia fetida*, *Eisenia hortensis* and *Lumbricus terrestris*) [62], they observed the accumulation of PCBs in all earthworm tissues. In the same way [63], they investigated the capacity of *E. fetida* to accumulate doxin-like polychlorinated biphenyls (DL-PCBs). Using contaminated soils from Sweden [64], they showed the occurrence of PCDDs and PCDFs in *E. fetida* tissues resulting from exposure, both *in situ* and in laboratory conditions.

Most of the works on PBDEs (**Table 1**) have focused on the genus *Eisenia* and more specifically the specie *fetida*. Only two other genera have been recorded, namely *Lumbriculus* (*variegatus* and *rubellus*) and *Allolobophora caliginosa trapezoides* (**Table 1**). In a study investigating PCBs and PBDEs in the soil-earthworm-hedgehog food chain, the accumulation of these compounds in *Lumbricus rubellus* tissues was noted [50]. In East China, *E. fetida* and *Allolobophora caliginosa trapezoides* collected from an e-waste dismantling area showed the accumulation of PCBs, PCDDs, PCDFs and PBDEs in their tissues [49]. The results showed measurable and sometimes high

Species	Units	ΣPBDEs	BDE 28	BDE 47	BDE 99	BDE 100	BDE 153	BDE 183	BDE 209	Reference
Eisenia andrei	ng.g^{-1} DW	7.52–76.3 (21 congeners)	0.01–0.13	0.14–4.91	0.05–5.42	0.04–1.53	0.02–0.66	0.02–0.41	6.85–59.1	[43]
	BSAF	0.8–4.74	1.32–18.4	0.84–26.9	0.33–14.4	1.04–20.3	0.21–3.57	0.1–1.3	1.08–4.81	
Eisenia fetida	μg.g^{-1} lipid								1.04–865.6 *(adult)*	[44]
	BSAF								0.074–0.123 *(adult)*	
Eisenia fetida	μg.g^{-1} DW (max)			16	32			12	5	[45]
	BSAF* OECD spiked			5–12	4–10		0–4	1–3	0–1	
	BSAF*			3–9	3–6		1–3	ND	ND	
Eisenia fetida	BSAF*			5–12	4–9		2–4	0–3	0–1.5	[46]
Eisenia fetida	μg.kg^{-1} lipid	1000–65 000 (6 congeners)								[47]
	BSAF*			5–20	3–13	4–17				
Lumbriculus variegatus	ng.g^{-1} lipid (values obtained)		314	5685	2649	813	693	315	8980	[48]
	BSAF*		2.18	4.09	0.905	1.63	0.611	0.320	0.103	
Eisenia fetida	ng.g^{-1} DW	0.2–4.44 (14 congeners)								[49]
Allolbophora caliginosa trapezoides	ng.g^{-1} DW	0.56–2.13 (14 congeners)								

Species	Units	ΣPBDEs	BDE 28	BDE 47	BDE 99	BDE 100	BDE 153	BDE 183	BDE 209	Reference
Lumbricus rubellus	$ng.g^{-1}$ DW	0.08–0.2								[50]
	BSAF*	1.99–5.67								
Eisenia fetida	$ng.g^{-1}$ DW	0.537–67.56 (8 congeners)	0.003–0.024	0.095–0.854	0.041–1.145	0.011–0.659	0.005–0.203	0.003–0.095	0.239–65.161	[51]
	$\mu g.kg^{-1}$ lip	5.03–63.3 *(adult)* 13.7–160 *(juvenile)*								
	BSAF*	0.01–0.3 *(adult)* 0.02–7.92 *(juvenile)*								

BSAF, [earthworms DW]/ [soil DW]; BSAF, [earthworm lipid]/ [soil OC]; OECD, artificial soil used; DW, dry weight; OC, organic content; PBDE, polybrominated diphenyl ethers; BDE, brominated diphenyl ethers + IUPAC Nb.*

Table 1.
Polybrominated diphenyl ethers (PBDEs) content and factor of bioaccumulation (biota-to-soil accumulation factor (BSAF)) in several earthworm species. All values gave the min-max observed, except when indicated.

Species	Units	HBBz	TBBPA	HBCDDs			ΣPBBs (=52 + 101 + 153)	Reference
				α	β	ϒ		
Eisenia fetida	ng.g^{-1} DW (max)			2400	400	580		[52]
	BSAF			1.28–11.2	0.42–1.71	0.38–1.95		
Eisenia fetida	ng.g^{-1} DW			1500–3000	200–580	200–610		[53]
	BSAF			11–21	1.1–3.2	1.1–4.1		
Metaphire guillelmi	ng.g^{-1} DW			500–800	200–480	130–220		
	BSAF			2.5–5.5	1.1–2.7	0.9–1.6		
Eisenia fetida	BSAF*		14					[54]
Metaphire guillelmi	BSAF*		11					
Eisenia fetida	BSAF (*TBBPA + Cd*)		1.1–4.5					[55]
Metaphire guillelmi	BSAF (*TBBPA + Cd*)		6.6–7.5					
Eisenia fetida	BSAF			21.8	2.28	3.77		[56]
	BSAF*			2.58	0.27	0.44		
Metaphire guillelmi	BSAF			6.21	2.81	1.16		
	BSAF*			1.10	0.50	0.20		
Eisenia fetida	μg.g^{-1} DW (max)	10						[45]
	BSAF* *OECD spiked*	2–6						
	BSAF*	2.6–8.9						
Eisenia fetida	ng.g^{-1} DW	0.21–0.44		0.027–4.68	0.067–0.095	0.045–1.970	0.009–0.207	[51]
	ng.kg^{-1} lip						0.12 and 0.17 all other <LOD *(adult)* all<LOD *(juvenile)*	
	BSAF*						0–0.26 *(adult)* 0 *(juvenile)*	

BSAF, [earthworms DW]/ [soil DW]; BSAF, [earthworm lipid]/ [soil OC]; OECD, artificial OECD soil; DW, dry weight; OC, organic content; [PBB, polybrominated biphenyls; HBBz, hexabromobenzene; TBBPA, tetrabromobisphenol A;TBECH, 2-dibromo-4-(1,2-dibromoethyl)cyclohexane; HBCDDs, hexabromocyclododecane]. Limit of detection (LOD) (PBBs), 0.034.*

Table 2.
Content and bioaccumulation factors of some brominated flame retardants (BFRs) in several earthworm species. All values gave the min-max observed except when indicated or when just one value was available.

		PCDD/Fs		PCBs			Legacy BFRs			e-BFRs						
		ΣPCDDs	**ΣPCDFs**	**Σ DL-PCBs-no**	**Σ DL-PCBs-mdio**	**Σ_6 NDL-PCBs**	**Σ_8 PBDEs**	**Σ HBCCD**	**Σ 3 PBBs**	**PBEB**	**PBT**	**TBCT**	**HBBz**	**PBBz**	**pTBX**	**OBIND**
Adult	Min	94.30	60.7	2390	24 0420	1 988 320	4.393	*<0.31*	*<0.034*	< *0.029*	0.058	< *0.032*	0.289	0.058	< *0.037*	*<0.178*
	Max	11 310.9	2 377.9	36 930	2 539 830	20 087 510	63.29	4.96	0.173	0.116	4.913	< *0.032*	4.393	1.098	< *0.037*	*<0.178*
Juvenile	Min	109.02	107.7	960	52 260	380 670	13.71	*<0.31*	*<0.034*	< *0.029*	0.337	< *0.032*	0.421	0.084	< *0.037*	*<0.178*
	Max	7 653.3	1 219	11 330	843 960	6 077 920	159.33	4.85	*<0.034*	0.337	4.972	< *0.032*	8.12	1.798	< *0.037*	*<0.178*

[PCDDs, polychlorinated dibenzo-p-dioxins; PCDFs, polychlorinated dibenzofurans; DL-PCBs-no, dioxin-like none ortho polychlorinated biphenyls; DL-PCBs-mdio, dioxin-like mono + di-ortho polychlorinated biphenyls; NDL-PCBs, non-dioxin-like polychlorinated biphenyls; PBDE, polybrominated diphenyl ethers; PBB, polybrominated biphenyls; PBEB, pentabromoethylbenzene; PBT, pentabromotoluene; TBCT, tetrabromo-o-chlorotoluene; HBBz, hexabromobenzene; PBBz, pentabromobenzene; pTBX, 2,3,5,6-Tetrabromo-p-xylene; OBIND, octabromotrimethylphenylindane]. Values indicated in italic with < symbol are the limit of quantitation (LOQ) for the compound.

Table 3.
Concentrations of several organic contaminants measured in Eisenia fetida *adults and juveniles exposed during 4 weeks on eight different soils (all results in ng.kg^{-1} lipid dw) [51].*

concentrations (of the order of μg.kg^{-1} lw), demonstrating the transfer of PBDEs to earthworms (**Table 1**). In addition, it was noted that some biota-to-soil accumulation factors (BSAFs) for PBDEs are very high, and not only for the most abundant congener in soils (BDE 209). These factors are just as important for other congeners such as BDE 47 (**Table 1**). The predominance of BDE 47 and BDE 99 is typical for biota [65, 66]. BDE 209 displayed a strong sorption tendency to soils/sediments and biological samples due to its extremely hydrophobic character (the log K_{ow} for BDE 209 is 9.97) [67]. Generally, PBDE profiles in environmental samples were dominated by BDE 209 [18]. High concentrations of PBDEs (average Σ_{10}PBDEs 2665 ng.g^{-1} dw) were detected in residential soils from an electronic waste polluted region in Guiyu, China, where BDE 209 accounted for 73% of the total Σ_{10}PBDEs [68]. *E. andrei* showed concentrations ranging over a factor of 10 between the minimum and maximum levels, from 7.5 to 75 ng.g^{-1} dw, with a predominance of BDE 209 in the tissues. On the other hand, comparing BSAFs, BDE 47 expressed the highest global factor (26.9), which reduced to 4.74 for ΣPBDEs (**Table 1**). In an experiment where earthworms were exposed to a low spiked artificial soil, *E. fetida* showed ΣPBDE concentrations of up to 65,000 ug.kg^{-1} lw [47]. Other studies showed ranges of *E. fetida* concentrations between 5 and 160 ug.kg^{-1} lw and 1 and 865.6 ug.g^{-1} lw (**Table 1**). The maximum BSAF values were 20 for BDE 47, 17 for BDE 100 and up to 7.92 for ΣPBDEs (**Table 1**). Tissue burdens decreased in the following order: BDE 47 > 99 > 100 > 85 > 154 > 153 [47]. In the present study, adult *E. fetida* showed a BSAF <1 for BDE 209, as observed in [44], and the highest values of up to 7.92 were seen in the tissues of juveniles [51] (**Table 1**). The *Lumbriculus* and *lumbricus* genera showed concentrations of up to 8900 ng.g^{-1} lw for BDE 209 and some BSAF values were showed up to 4.09 for BDE 47 and up to 5.67 for ΣPBDEs (**Table 1**). In a study on Canadian (British Columbia) soils [69], they reported BSAF (normalized to lipid in biota and organic content (OC) in soil) values of 4.74 for BDE 47, 33 for BDE 85, 2.56 for BDE 99 and 2.47 for BDE 100, for the "group" *lumbricidae*. Some field studies investigated the accumulation of PBDEs in earthworms, and expressed the relationship between BSAFs and the octanol–water partition coefficient (log K_{ow}), to assess bioavailability to earthworms, and did not address the direct relationship between soil and earthworm concentrations [70, 71].

However, positive relationships between soil and earthworm (*Lumbricus rubellus)* contents were found for PBDEs (sum of two congeners 47 and 99) for the combined data for two parks where hedgehogs foraged [50].

Concerning the other legacy BFRs (**Table 2**), we noted that only two types of earthworms, *Eisenia fetida* and *Metaphire guillelmi,* were used to study soil interactions. Several studies have shown the transfer of HBCDDs to earthworms (**Table 2**). In studies with *E. fetida*, the highest concentrations were observed for α-HBCDD, showing a range from 2400 to 3000 ng.g^{-1} dw. Corresponding levels were 10 times lower for β-HBCDD with values between 200 and 400 ng.g^{-1} dw, with similar concentrations for γ-HBCDD ranging from 200 to 610 ng.g^{-1} dw (**Table 2**). BSAFs reflected the dominance of α-HBCDD with a maximum value of ∼22 and a range from 1 or 2, up to 21, in comparison to the lower factors for β-HBCDD and γ-HBCDD, ranging from <1 to 4.1 (**Table 2**). Similarly, in studies on *M. guillelmi*, the highest concentrations at 500 and 800 ng.g^{-1} dw were noted for α-HBCDD, one third of the maximal observed values for *E. fetida* for the same congener (**Table 2**). However, the concentration ranges for β-HBCDD (200 and 480 ng.g^{-1} dw) and γ-HBCDD (130 and 220 ng.g^{-1} dw) were similar to the corresponding ranges seen in *E. fetida* (**Table 2**). Such transfer can be explained by the highly hydrophobic character of individual

HBCDDs, with $\log K_{ow}$ values of 5.38 (α-), 5.47 (β-) and 5.80 (γ-), and their association, primarily with soil particles [72]. BSAFs reflect this dominance of α-HBCDD in *M. guillelmi* with a maximum α-HBCDD factor at 5.5 in comparison to β-HBCDD and γ-HBCDD factors, with maximum BSAF values of 2.7 and 1.6, respectively (**Table 2**). Also, reflecting the relatively lower concentrations, *M. guillelmi* expressed a lower transfer factor than *E. fetida,* with BSAFs at least two to four times lower (**Table 2**). This observed difference in the bioaccumulation of β-HBCDD between the two species can be due to other processes, such as uptake, depuration, metabolism and isomerization. In earthworms, β- and γ-HBCDDs are strongly bioisomerized to the α-form, but this isomerization occurs to a larger extent in *E. fetida* [56]. The difference in stereostructure of the three HBCDD diastereoisomers leads to substantial differences in their physicochemical, biological and toxicological properties [73, 74]. The lipids in an organism are an important depot for the storage of organic pollutants. In Ref. [52], they report the lipid content of the body wall and gut at higher than 5%, underlining the important potential role that the tissue lipid plays in affecting the accumulation and distribution of HBCDDs in earthworms, compared to protein. In comparison, *Lumbricus rubellus* showed a lipid content of between 6.16 and 6.96% lipid weight (lw)/dry weight (dw) basis [50]. In Ref. [56], they explained the much greater BSAF values of α- and γ-HBCDDs in *Eisenia fetida* relative to *Metaphire guillelmi*, by the higher lipid and protein contents of tissues in *E. fetida*. Some studies have reported that the relative accumulation capacity of animal protein was approximately 5% that of lipid. Therefore, in animals with lipids at <5% of the dry weight content, protein sorption will exert a significant role in the accumulation and distribution of contaminants [75, 76]. The uptake in earthworms of other BFRs such as TBBPA has also been investigated (**Table 2**). TBBPA was studied in combination with cadmium [55] (**Table 2**), with results for TBBPA showing a BSAF value between 1.1 and 14 for *E. fetida* and between 6.6 and 14 for *M. guillelmi*. **Table 2** also shows studies on HBBz, uptake in *E. fetida*, with concentrations of 0.021–10,000 $ng.g^{-1}$ dw and BSAF values between 2 and 8.9. In our studies on ΣPBBs, we noted low transfer values from soil to earthworms for the eight samples tested, with only two quantifiable values of 120 and 170 $ng.kg^{-1}$ lw in adults. All other concentrations were below the limit of detection (LOD), including in all juvenile earthworms [51] (**Table 2**). The three PBBs (32, 101 and 153) investigated, seemed to indicate little, if any transfer to the earthworms, particularly in juveniles. The lack of other studies on PBB uptake in earthworms limits the possible comparisons and discussion.

If we compare the e-BFR concentrations in *Eisenia fetida* to values obtained for other POPs such as PCBs and PCDD/Fs (**Table 3**), we also notice the concentration differences between adults and juveniles exposed to eight different soils during 28 days [51].

For adults, we noted concentrations decreasing in the following order: NDL-PCBs (non-dioxin-like polychlorinated biphenyls) >> > mono-di-ortho-PCBs » non-ortho-PCBs>>>PCDDs>>PCDFs>PBDEs>>HBCCD≈HBBz≈PBT>PBBz>PBBs≈PBEB>>-LOQ (limit of quantitation): TBCT (tetrabromo-o-chlorotoluene); pTBX (2,3,5,6-tetrabromo-p-xylene); OBIND (octabromotrimethylphenylindane). For juveniles, we noted small differences in this order: NDL-PCBs >> > mono-di-ortho-PCBs » non-ortho-PCBs>>>PCDDs>>PCDFs>PBDEs>>HBBz>HBCCD≈≈PBT>>PBBz>>PBEB>>L-OQ: TBCT; pTBX; OBIND (**Table 3**). The preferential accumulation of some POPs in earthworms, especially NDL-PCBs and DL-PCBs, was expected in this species, following a study by [63], on the capacity of *E. fetida* to DL-PCBs.

In a study with four different species (*Eisenia andrei*, *Eisenia fetida*, *Eisenia hortensis* and *Lumbricus terrestris*), the accumulation of PCBs was observed in all earthworm tissues [62]. In Japan, a study of earthworms in rice fields showed tissue levels of 150 μg.kg^{-1} fresh weight of DL-PCBs [77]. In East China, *E. fetida* and *Allolobophora caliginosa trapezoides* species from a typical e-waste dismantling area showed PCB accumulation in tissues at levels of 1.17 up to 78.6 μg.kg^{-1} dw [50]. In Ref. [50], they also observed that earthworm PCB concentrations increased proportionately with increasing soil levels of PCBs. However, BSAFs decreased significantly with increasing soil concentrations of dichlorodiphenyltrichloroethane (DDT) ($r = -0.52$, $p = 0.005$), PCBs ($r = -0.89$, $p = 0.001$) and PBDEs ($r = -0.85$, $p = 0.01$) suggesting that steady-state equilibrium of compound distribution was not reached in soils and worms in this field study [50]. In Ref. [64], they also demonstrated, both *in situ* and during exposure in laboratory conditions, the accumulation of PCDDs and PCDFs in *E. fetida* tissues at concentrations of 1.5–15,000 μg.kg^{-1} dw, using contaminated soils from Sweden. The accumulation of PCDDs and PCDFs in the tissues of two other earthworm species, *Allolobophora caliginosa trapezoides* and *Lumbricus rubellus*, was also observed by [78], and in [77] they reported concentrations of 0.9 μg.kg^{-1} dw of PCDDs and PCDFs in earthworms taken from rice fields in Japan. In another study [49], PCDD and PCDF accumulation at a range between 0.13 and 0.59 μg.kg^{-1} dw was recorded in *E. fetida* and *Allolobophora caliginosa trapezoides* species collected from a typical e-waste dismantling area (in East China). Using radio labeled ^{14}C-TBBPA, they showed an uptake (ks) of this BFR at 0.03 (± 0.002) $g_{oc}.g^{-1}$ lw.day^{-1} for *E. fetida* and 0.08 (± 0.03) $g_{oc}.g^{-1}$ lw.day^{-1} for *M. guillelmi* [54].

Considering the BFRs measured, we noted that while PBDEs were frequently detected, some e-BFRs, such as HBBz and PBBz, were also efficiently bioaccumulated (**Tables 2** and **3**). Some BSAF values presented (**Tables 1–3**) clearly show the potential of BFRs to accumulate in *E. fetida* tissues. In our results, adult and juvenile *E. fetida* can be considered as poor bioaccumulators of PCDDs, PCDFs, PBDEs and PBBs (bioaccumulation factor (BAF) < 1) but show strong bioaccumulation of PCBs, HBBz, PBBz, PBT and PBEB (BAF > 2) (**Table 3**) [51]. Moreover, in our studies on *E. fetida*, bioaccumulation factors (BAFs) for some of the studied BFRs were lower than 10, indicating the potential transfer of these POPs to higher trophic levels along with the potential toxicity effects. The data show that BFRs are present in the soils at much lower levels than PCBs, PCDDs and PCDFs, but despite this, they are also available for incorporation in soil-dwelling organisms and possible biomagnification through upper levels of food chains.

Both brominated and chlorinated compounds undergo the same processes of volatilization in the atmosphere, and of dissolution in water leading to accumulation by aquatic organisms and adsorption to soil and/or sediment particles. This fact supports the hypothesis that some brominated compounds (notably BDE 153, BDE 99, BDE 47 and BDE 28) have a similar environmental behavior to chlorinated compounds such as PCBs, PCDDs and PCDFs.

2.3 Effects of BFRs on earthworms

Earthworms are more and more recognized as an important model test organism for studying the effects of contaminants on soil organisms, notably for establishing standard guidelines for risk evaluation. First, the European Union (EU) and the Organization for Economic Cooperation and Development (OECD) defined acute toxicity tests for earthworms based on mortality and growth rate [79] and also an avoidance test [78, 80]. Later, OECD proposed chronic toxicity assays based on reproduction rate measurement [81]. Most of these studies were done with legacy BFRs, usually PBDEs. Moreover, in order to normalize the tests, as well as for reasons of reproducibility and "simple" breeding, the most widely used genus is *Eisenia* and particularly *Eisenia fetida*, as a model organism.

2.3.1 Global toxicological effects and distribution in the organism

Most studies on bioaccumulation from soils have been done on PBDEs, notably BDE 209. Thus, avoidance behavior studies show that *E. fetida* was insensitive to BDE 209 exposure for 48 h, at concentrations between 0.1 and 1000 $mg.kg^{-1}$ [44], or to a 231–3720 $mg.kg^{-1}$ decabromodiphenyl ethane exposure [82]. The European Commission [83] also observed that no abnormal avoidance behaviors occurred when *E. fetida* was exposed to BDE 209.

No significant effects of BDE 209 on earthworm cocoon production were detected but a statistically significant decrease in the number of juveniles per hatched cocoon was reported when artificial soils were spiked at 1000 $mg.kg^{-1}$ with BDE 209 [44]. In Ref. [82], they also reported no differences between decabromodiphenyl ethane exposure and the control, in terms of percent mortality, mass and weight gain for the adult earthworms. But they did report adverse effects on reproduction at a high dose (3720 $mg.kg^{-1}$ dw [82]).

Concerning toxicological endpoints, the NOEC (no observed effect concentration) and LOEC (lowest observed effect concentration) for 56-day reproduction were set, respectively, at 1910 and 3720 $mg.kg^{-1}$ dw for decabromodiphenyl ethane (DBDPE) for *Eisenia fetida* [82]. The lowest observed effect level (LOEL) for the parameters of survival and growth in *E. fetida* was reported at more than 1000 $mg.kg^{1}$ of BDE 209 [44].

This is a high level and one hypothesis is that adult earthworms are more tolerant to BDE 209 exposure in soils, unlike juveniles, that seem to be more prone to the potential toxicity [44].

Studies on the HBCDD tissue distribution in *E. fetida* found a gradient order of concentration as: gut>body fluid>body wall [52, 84]. The report on the lower concentration of HBCDDs in earthworm's casts relative to the paired bulk soil, observed in [52], allows us to conclude that earthworms take up HBCDDs mainly through digestion of soil particles. This was confirmed by the fact that the highest HBCDD concentrations and BAFs were found in the gut [52]. Thus, HBCDD accumulated in the gut was translocated into the body fluid and distributed to the body wall notably by a series of passive diffusion and partition processes. Moreover, more than 5% of lipid content was measured in the body wall and gut, indicating that tissue lipids can play an important role in the HBCDDs' accumulation and distribution in earthworms compared to protein [52].

2.3.2 Metabolic and physiological effects

The metabolic responses of earthworms to BFRs appear to be more sensitive than growth inhibition and/or the reproduction global test. PBDEs were reported to increase the intracellular reactive oxygen species (ROS) level and then induce oxidative stress with some effects on the activities of enzymes, such as catalase (CAT), glutathione-S-transferase (GST), superoxide dismutase (SOD) or on the level of malondialdehyde (MDA) [85].

In Ref. [86], enzyme activity modifications in earthworms, which were in contact with soils treated with BDE 209, relative to the controls, showed that: (i) superoxide dismutase (SOD) activities were elevated significantly after 21 and 28 days of exposure; (ii) catalase (CAT) activities were much higher in all tests during the entire exposure period; (iii) peroxidase (POD) and glutathione-S-transferase (GST) activities generally decreased and indicated a contrary response trend; and (iv) total antioxidant capacity (T-AOC) after exposure to a low level of BDE 209 (1 $mg.kg^{-1}$) was induced, whereas it was suppressed at 10 and 100 $mg.kg^{-1}$.

In Ref. [85], they showed a slight induction of oxidative stress for *Eisenia fetida* caused by lower levels of BDE 209, and severe oxidative stress at higher levels. The electron paramagnetic resonance (EPR) spectra suggested that hydroxyl radicals (•OH) in earthworms were significantly induced by BDE 209, and that the changes in MDA content indicated that reactive oxygen species (ROS) might lead to cellular lipid peroxidation in earthworms [86]. These results demonstrated that BDE 209 toxicity plays an important role in inducing severe oxidative stress in *E. fetida* [86].

Another study on *E. fetida* exposed to sublethal concentrations of BDE 47 and 209 [87] demonstrated that there is an increase in protein degradation to amino acids (with lactate and glutamate accumulation) in earthworms. The increase in the betaine content indicated that *E. fetida* may maintain cell osmotic pressure and protect enzyme activity by metabolic regulation [87]. The exposure interval and concentration of contaminants also showed a significant effect on the lysosomal stability of earthworms [88]. Earlier studies had shown a decrease in the lysosomal stability of *Eisenia fetida* [89, 90]. These reported observations clearly demonstrate that PBDEs induce changes in the enzyme activities, causing oxidative stress as well as disruption to the metabolic activities of *Eisenia fetida*. A recent study carried out using artificial soil fortified with different concentrations of DBDPE also showed biochemical toxicity in *Eisenia fetida* [91]. This e-BFR caused an increase in ROS and MDA levels as well as some modifications in the levels of antioxidant enzymes [91]. The authors showed DNA damage at all tested DBDPE concentrations, additionally suggesting a genotoxic response [91].

2.3.3 Kinetics of absorption/depuration

Polybrominated diphenyl ethers in soil become less bioavailable when their $\log K_{ow}$ increases [43]. The limited bioavailability of some PBDE congeners can be due to their large molecular size (>0.95 nm), high molecular weight (959.17 $g.mol^{-1}$ for BDE 209) and the high hydrophobicity of these compounds [43]. Such properties limit the ability of PBDEs to penetrate cell membranes and decrease the uptake efficiency by earthworms. But in Ref. [88], they demonstrated the effects of earthworms on the removal of BDE 209 in contaminated soils, demonstrating the decrease with kinetic curve plots of the decline in BDE 209 soil content. They also observed that this process followed a normal biphasic pattern *i.e.* a fast-initial decrease during the first few days, followed by a slower phase with a smooth decrease until the end of exposure [88]. A similar two-step elimination was also observed in an earlier study on other terrestrial organisms such as *Enchytraeus albidus* and *Enchytraeus luxuriosus* exposed to lindane and HCB [92]. BFR concentrations in the oligochaete, *Lumbriculus variegatus*, also increased rapidly in the 14 first days before arriving at a "plateau" toward the end of 28 days of exposure [48].

Earthworms take up organic contaminants through two main pathways, passive epidermal uptake and dietary uptake. It has been reported that the contributions of the two routes are different, depending on the properties of the contaminant. Dietary uptake is the dominant route for hydrophobic compounds that have a logK_{ow} value > 5 [93, 94]. HBCDDs are highly hydrophobic, with logK_{ow} values of 5.38–5.80, and are primarily associated with soil particles [72]. As logK_{ow} values for PBDEs were between 5.9 and 10, dietary uptake is also likely to be the dominant exposure route for these compounds in earthworms. The lipid contents of earthworms are also an important parameter for the accumulation of organic pollutants. In fact, the greater the lipid and protein contents in *E. fetida*, the higher the bioaccumulation potential [56], but several different processes (such as uptake, depuration, metabolism and isomerization) also play a role in the final accumulation of contaminants. For example, *E. fetida* showed 1.73% wt dw and 3.56% wt dw of lipids, respectively, for adults and juvenile earthworms, predicting differences in the uptake potential/capacity between adults and juveniles [51].

In Ref. [88], they observed that BDE 209 was likely to lose one bromine atom in the biota. The proportions of lower brominated congeners (BDE 153, BDE 99, BDE 47 and BDE 28) in earthworms (34.5–52.5%) were generally higher than in soil (27.9–40.5%), indicating either that the debromination mechanism occurred more easily in earthworms or that BDE 153, BDE 99, BDE 47 and BDE 28 were easily absorbed by earthworms with a low elimination rate [88].

The depuration rates for the three HBCDD diastereomers in two earthworm species, *E. fetida* and *Metaphire guillelmi,* followed the order β- > γ- > α- [56]. β- and γ–HBCDDs were bioisomerized to α-HBCDDs, to a greater extent in *E. fetida* than in *M. guillelmi*. The depuration half-time values ($t_{1/2}$) were greater for α-HCBDDs (30.1–12.8 days) than for γ HBCDD (14.7–6.86) or β-HBCDD (7.88–5.82) [56].

The BFR, TBBPA and cadmium (Cd) are three ubiquitous pollutants in soils and are often found together, notably in electronic waste recycling sites. In an experimental study [55], they investigated the toxicity of these pollutants, as well as the accumulation and subcellular partitioning in two earthworms species, *Eisenia fetida* and *Metaphire guillelmi,* exposed for 14 days to Cd (at 1 mg.kg^{-1}) and TBBPA (at 10, 50, 100 and 500 mg.kg^{-1}) separately and in combination. In general, the Cd-TBBPA co-exposure resulted in synergistic effects in terms of acute toxicity, growth inhibition, histopathological changes in body walls and oxidative stress responses to earthworms [55]. Moreover, *M. guillelmi* showed higher sensitivity to Cd-TBBPA co-exposure and a higher Cd BSAF than *E. fetida,* [55]. By contrast, the increased Cd accumulation in *M. guillelmi* suggested a different uptake route than that of *E. fetida*, specifically gut processes and epidermal absorption, respectively [55]. The study showed that, in addition to specific pathways or lipid contents, other interactions between pollutants can modify the preferential transfer route of POPs in earthworms.

2.4 Conclusions on BFRs in earthworms

Earthworms can be considered as valuable bioindicators of soil pollution as they are very sensitive to contaminants in the soil and are significantly affected by them. Earthworms are exposed to soil contaminants through different routes of exposure such as direct contact with soil pore water or through their food. Their position at the base of the food chain and their abilities to uptake POPs make them an important route for pollutant entry into the terrestrial ecosystem and biomagnification to higher trophic levels. Several studies have already demonstrated the accumulation of POPs, such as PCBs, PCDD/Fs and PBDEs, in earthworm tissues at considerable levels without causing mortality or severe toxic effects. Our study is one of the few that show the transfer and bioaccumulation of e-BFRs in earthworms, and the differences in effects between adult and juvenile *E. fetida*.

Earthworms bioaccumulate BFRs passively through dermal absorption, and actively through ingestion. Effectively, the dietary pathway to bioaccumulation is facilitated by the lipid content of the earthworms' gut and body wall, and is also influenced by the hydrophobicity of the contaminants. But the phenomenon of bioaccumulation is very complex, with a species specificity (for dermal composition, metabolization and/or depuration, behavior) that is controlled by physicochemical properties of both the contaminants and the soil: the concentration and speciation of the contaminants, the type and characteristics of the soil, the temperature, the duration of exposure, the bioaccessibility/mobility of the contaminants and their interactions with the other contaminants present in the soil.

3. General conclusion and future directions

The data on BFR levels and trends in soil need to be extended, notably for e-BFRs, for which little data exist. The levels of these contaminants in soils are currently low but measurable. The time trends of BFRs showed that legacy BFRs, such as the well-studied PBDEs, were increasing in terrestrial environments and accumulating in wildlife, notably in earthworms. A study of the differences in the environmental distributions between legacy BFRs and e-RFBs seems to be, for now at least, hindered by the lack of occurrence data that would have allowed dispersion modeling and/or bioaccumulation studies. This makes us wonder whether in the near future, e-BFRs will overtake the levels and dispersion in ecosystems of legacy BFRs such as PBDEs for example. In the same way as what has been studied for PCBs, we should investigate if the persistence, or maybe the decrease of legacy BFRs levels in soils, could be tracked in relation to the source-limiting effects of political and regulatory decisions to no longer use these chemicals.

The risk evaluation for food chain linked to bioaccumulation remains to be investigated. Bioaccumulation or bioconcentration factors (BAF/BCF), which have been studied for legacy BFRs, should now be performed for e-BFRs. In risk assessments, BAFs are used to estimate and predict the potential trophic transfer of contaminants from soil to wildlife. Therefore, in order to obtain a better understanding of the environmental fate of e-BFRs, their accumulation, dispersion, or in order to model their fluxes, further studies would be useful and should be encouraged. Due to the increasing levels of these organic compounds in terrestrial ecosystems, it is important to study the occurrence, fate and transfer processes of BFRs in earthworms, as well as the potential phenomenon of trophic biomagnification. These studies are essential to evaluate and manage the risks posed by organic pollutants, such as BFRs or e-BFRs, to ecosystems and human health.

Acknowledgements

We are grateful to Alwyn R. Fernandes from the University of East Anglia for helping us to improve the quality of the manuscript and for his valuable comments. This work was funded by the LabEx DRIIHM, French program "Investissements d'Avenir" (ANR-11-LABX-0010) managed by the ANR. The thesis of Claudia Coelho was financed by the LabEx DRIIHM and was also supported by the OHM Vallée du Rhône (France) and OHMi Estarreja (Portugal).

Conflict of interest

The authors declare no conflict of interest.

Author details

Jean-Philippe Bedell[1]*, Claudia Coelho[1], Olivier Roques[1], Anais Venisseau[2], Philippe Marchand[2] and Yves Perrodin[1]

1 University of Lyon, Université Claude Bernard Lyon 1, CNRS, ENTPE, UMR 5023, LEHNA, Vaulx-en-Velin, France

2 ONIRIS, LABERCA, Nantes, France

*Address all correspondence to: bedell@entpe.fr

References

[1] Birnbaum LS, Staskal DF. Brominated flame retardants: Cause for concern? Environmental Health Perspectives. 2004;**112**(1):9-17. DOI: 10.1289/ehp.6559

[2] de Wit CA. An overview of brominated flame retardants in the environment. Chemosphere. 2002; **46**(5):583-624. DOI: 10.1016/s0045-6535(01)00225-9

[3] Alaee M, Arias P, Sjödin A, Bergman Å. An overview of commercially used brominated flame retardants, their application, their use pattern in different countries/regions and possible modes of release. Environment International. 2003;**29**(6):683-689. DOI: 10.1016/S0160-4120(03)00121-1

[4] de Wit CA, Herze D, Vorkamp K. Brominated flame retardants in the artic environment–Trends and new candidates. Science of the Total Environment. 2010;**408**(15):2885-2918. DOI: 10.1016/j.scitotenv.2009.08.037

[5] Covaci A, Harrad S, Abdallah MAE, Ali N, Law RJ, Herzke D, et al. Novel brominated flame retardants: A review of their analysis, environmental fate and behaviour. Environment International. 2011;**37**:532-556. DOI: 10.1016/j.envint.2010.11.007

[6] Bergman A, Ryden A, Law RJ, de Boer J, Covaci A, Alaee M, et al. A novel abbreviation standard for organobromine, organochlorine and organophosphorus flame retardants and some characteristics of the chemicals. Environment International. 2012;**49**: 57-82. DOI: 10.1016/j.envint.2012.08.003

[7] EFSA Panel on Contaminants in the Food Chain (CONTAM). Scientific opinion on emerging and novel brominated flame retardants (BFRs) in food. EFSA Journal. 2012;**10**(10):2908. [133 p.]. DOI: 10.2903/j.efsa.2012.2908

[8] Ezechiáš M, Covino S, Cajthaml T. Ecotoxicity and biodegradability of new brominated flame retardants: A review. Ecotoxicology and Environmental Safety. 2014;**110**:153-167. DOI: 10.1016/j.ecoenv.2014.08.030

[9] Linares V, Bellés M, Domingo JL. Human exposure to PBDE and critical evaluation of health hazards. Archives of Toxicology. 2015;**89**(3):335-356. DOI: 10.1007/s00204-015-1457-1

[10] Loganathan BG. Global contamination trends of persistent organic chemicals: An overview. In: Loganathan B, Lam PKS, editors. Global Contamination Trends of Persistent Organic Chemicals. Boca Raton, FL: CRC Press; 2012. pp. 3-32

[11] Vastag B. EPA feels heat over flame retardant. Nature. 2008;**452**:513. DOI: 10.1038/452513a

[12] Legler J, Brower A. Are brominated flame retardants endocrine disruptors? Environment International. 2003;**29**(6): 879-885. DOI: 10.1016/S0160-4120(03)00104-1

[13] Cao Z, Yu G, Chen Y, Liu C, Liu K, Zhang T, et al. Mechanisms influencing the BFR distribution patterns in office dust and implications for estimating human exposure. Journal of Hazardous Materials. 2013;**52**:11-18. DOI: 10.1016/j.jhazmat.2013.02.043

[14] McGrath TJ, Morrison PD, Sandiford CJ, Ball AS, Clarke BO. Widespread polybrominated diphenyl ether (PBDE) contamination of urban soils in Melbourne, Australia.

Chemosphere. 2016;**164**:225-232. DOI: 10.1016/j.chemosphere. 2016.08.017

[15] Liu K, Li J, Yan SJ, Zhang W, Li YJ, Han D. A review of status of tetrabromobisphenol A (TBBPA) in China. Chemosphere. 2016;**148**:8-20. DOI: 10.1016/j.chemosphere. 2016.01.023

[16] Lu J-F, He M-J, Yang Z-H, Wei S-Q. Occurrence of tetrabromobisphenol A (TBBPA) and hexabromocyclododecane (HBCD) in soil and road dust in Chongqing, western China, with emphasis on diastereoisomer profiles, particle size distribution, and human exposure. Environmental Pollution. 2018;**242**:219-228. DOI: 10.1016/j. envpol.2018.06.087

[17] McGrath TJ, Morrison PD, Ball AS, Clarke BO. Detection of novel brominated flame retardants (E-BFRE-BFRs) in the urban soils of Melbourne, Australia. Emerging Contaminants. 2017;**3**:23-31. DOI: 10.1016/j. emcon.2017.01.002

[18] Hites RA. Polybrominated diphenyl ethers in the environment and in people: A meta-analysis of concentrations. Environmental Science & Technology. 2004;**38**(4):945-956. DOI: 10.1021/ es035082g

[19] Hao Y, Meng W, Li Y, Han X, Lu H, Wang P, et al. Concentrations and distribution of novel brominated flame retardants in the atmosphere and soil of Ny-Ålesund and London Island, Svalbard. Journal of Environmental Sciences. 2020;**97**:180-185. DOI: 10.1016/j.jes.2020.04.031

[20] Wang S, Wang Y, Song M, Luo C, Li J, Zhang G. Distributions and compositions of old and emerging flame retardants in the rhizosphere and non-rhizosphere soil in an e-waste contaminated area of South China. Environmental Pollution. 2016;**208**: 619-625. DOI: 10.1016/j. envpol.2015.10.038

[21] Zheng Q, Nizzetto L, Li J, Mulder MD, Sáňka O, Lammel G, et al. Spatial distribution of old and emerging flame retardants in Chinese forest soils: Sources, trends and processes. Environmental Science & Technology. 2015;**49**:2904-2911. DOI: 10.1021/ es505876k

[22] Matsukami H, Suzuki G, Someya M, Uchida N, Tue NM, Tuyen LH, et al. Concentrations of polybrominated diphenyl ethers and alternative flame retardants in surface soils and river sediments from an electronic waste processing area in northern Vietnam, 2012–2014. Chemosphere. 2017;**167**: 291-299. DOI: 10.1016/j. chemosphere.2016.09.147

[23] Someya M, Suzuki G, Ionas AC, Tue NM, Xu F, Matsukami H, et al. Occurrence of emerging flame retardants from e-waste recycling activities in the northern part of Vietnam. Emerging Contaminants. 2016; **2**:58-65. DOI: 10.1016/j. scitotenv.2015.02.008

[24] Wei YL, Bao LJ, Wu CC, Zeng EY. Characterization of anthropogenic impacts in a large urban center by examining the spatial distribution of halogenated flame retardants. Environmental Pollution. 2016;**215**: 187-194. DOI: 10.1016/j. envpol.2016.05.009

[25] Hong W-J, Jia H, Ding Y, Li W-L, Li Y-F. Polychlorinated biphenyls (PCBs) and halogenated flame retardants (HFRs) in multi-matrices from an electronic waste (e-waste) recycling site in northern China. Journal

of Material Cycles and Waste Management. 2016;**20**:80-90. DOI: 10.1007/s10163-016-0550-8

[26] Li WL, Liu LY, Zhang ZF, Song WW, Huo CY, Qiao LN, et al. Brominated flame retardants in the surrounding soil of two manufacturing plants in China: Occurrence, composition profiles and spatial distribution. Environmental Pollution. 2016;**213**:1-7. DOI: 10.1016/j.envpol.2016.01.092

[27] Wu JP, Zhang Y, Luo XJ, She Y, Yu L, Zhang Y, et al. A review of polybrominated diphenyl ethers and alternative brominated flame retardants in wildlife from China: Levels, trends, and bioaccumulation characteristics. Journal of Environmental Sciences. 2012; **24**(2):183-194. DOI: 10.1016/s1001-0742(11)60758-4

[28] Kammenga JE, Dallinger R, Donker MH, Köhler HR, Simonsen V, Triebskorn R, et al. Biomarkers in terrestrial invertebrates for ecotoxicological soil risk assessment. Reviews of Environmental Contamination and Toxicology. 2000; **164**:93-147

[29] Edwards CA. The importance of earthworms as key representatives of the soil fauna. In: Edwards CA, editor. Earthworm Ecology. 2nd ed. Boca Raton, FL: CRC Press; 2004. pp. 3-11. DOI: 10.1201/9781420039719

[30] Barlett MD, Briones MJI, Neilson R, Schmidt O, Spurgeon D, Creamer RE. A critical review of current methods in earthworm ecology: From individuals to populations. European Journal of Soil Biology. 2010;**46**:67-73. DOI: 10.1016/j.ejsobi.2009.11.006

[31] Lavelle P, Spain A. Soil Ecology. Amsterdam, Netherlands: Kluwer Scientific Publications; 2001. p. 654. DOI: 10.1007/978-94-017-5279-4

[32] Emmerling C, Paulsh D. Improvement of earthworm (Lumbricidae) community and activity in mine soils from open-cast coal mining by the application of different organic waste materials. Pedobiologia. 2001;**45**:396-407. DOI: 10.1078/0031-4056-00095

[33] Lavelle P, Barot S, Blouin M, Decaëns T, Jimenez J, Jouquet P. Earthworms as key actors in self-organized soil systems. In: Cuddington K, Byers JE, Wilson WG, Hastings A, editors. Theoretical Ecology Series. Vol. 4. Netherlands: Academic Press, Springer; 2007. pp. 77-106. DOI: 10.1016/S1875-306X(07)80007-4

[34] Bouché A. The establishment of earthworm communities. In: Satchell JE, editor. Earthworm Ecology: From Darwin to Vermiculture. London: Chapman and Hall; 1983. pp. 431-438. DOI: 10.1007/978-94-009-5965-1

[35] Bouché A. Earthworm species and ecotoxicological studies. In: Greig-Smith PW, Becker H, Edwards PJ, Heimbach F, editors. Ecotoxicology of Earthworms. Andover, UK: Intercept; 1992. pp. 22-35. DOI: 10.2134/jeq1994.00472442 50023000439x

[36] Hopkin SP. Ecophysiology of Metals in Terrestrial Invertebrates. Netherlands: Springer; 1989. p. 384

[37] Lanno R, Wells J, Conder J, Bradham K, Basta N. The bioavailability of chemicals in soil for earthworms. Ecotoxicology and Environmental Safety. 2004;**57**:39-47. DOI: 10.1016/j.ecoenv.2003.08.014

[38] Wallwork JA. Earthworm Biology. Coll. Studies in Biology, ed. London: Hodder Arnold; 1983. p. 64

[39] Jager T, Fleuren RHL, Hogendoorn EA, de Korte G. Elucidating the routes of exposure for organic chemicals in the earthworm, Eisenia andrei (Oligochaeta). Environmental Science & Technology. 2003;**37**: 3399-3404. DOI: 10.1021/es0340578

[40] Vijver MG, Vink JPM, Jager T, Wolterbeek HT, van Straalen NM, van Gestel CAM. Biphasic elimination and uptake kinetics of Zn and Cd in the earthworm Lumbricus rubellus exposed to contaminated floodplain soil. Soil Biology and Biochemistry. 2005;**37**: 1843-1851. DOI: 10.1016/j.soilbio.2005.02.016

[41] Morgan AJ, Stürzenbaun SR, Winters C, Grime GW, Aziz NAA, Kille P. Differential metallothionein expression in earthworm (Lumbricus rubellus) tissues. Ecotoxicology and Environmental Safety. 2004;**57**:11-19. DOI: 10.1016/j.ecoenv.2003.08.022

[42] Reinecke S, Reinecke A. Earthworms as test organisms in ecotoxicological assessment of toxicant impacts on ecosystems. In: Edwards CA, editor. Earthworm Ecology. 2nd ed. Boca Raton: CRC Press; 2004. p. 456. DOI: 10.1201/9781420039719.pt8

[43] Navarro I, de la Torre A, Sanz P, Pro J, Carbonell G, Martínez MA. Bioaccumulation of emerging organic compounds (perfluoroalkyl substances and halogenated flame retardants) by earthworms in biosolid amended soils. Environmental Research. 2016;**149**: 32-39. DOI: 10.1016/j.envres.2016.05.004

[44] Xie X, Qian Y, Wu Y, Yin J, Zhai J. Effect of decabromodiphenyl ether (BDE-209) on the avoidance response, survival, growth and reproduction of earthworms (Eisenia fetida). Ecotoxicology and Environmental Safety. 2013;**90**:21-27. DOI: 10.1016/j.ecoenv.2012.12.009

[45] Nyholm JR, Asamoah RK, Van Der Wal L, Danielsson C, Andersson PL. Accumulation of polybrominated diphenyl ethers, hexabromobenzene, and 1,2- dibromo-4-(1,2- dibromoethyl) cyclohexane in earthworm (Eisenia fetida). Effects of soil type and aging. Environmental Science & Technology. 2010;**44**:9189-9194. DOI: 10.1021/es1023288

[46] Nyholm JR. Persistency, bioaccumulation and toxicity assessment of selected brominated flame retardants [thesis]. Sweden: Umea University; 2009. p. 57

[47] Gaylor MO, Harvey E, Hale RC. Polybrominated diphenyl ether (PBDE) accumulation by earthworms (Eisenia fetida) exposed to biosolids-, polyurethane foam microparticle-, and penta-BDE amended soils. Environmental Science & Technology. 2013;**47**:13831-13839. DOI: 10.1021/es403750a

[48] Zhang B, Li H, Wei Y, You J. Bioaccumulation kinetics of polybrominated diphenyl ethers and decabromodiphenyl ethane from field-collected sediment in the oligochaete Lumbriculus variegatus. Environmental Toxicology and Chemistry. 2013;**32**(12): 2711-2718. DOI: 10.1002/etc.2384

[49] Shang H, Wang P, Wang T, Wang Y, Zhan H, Fu J, et al. Bioaccumulation of PCDD/Fs, PCBs and PBDEs by earthworms in field soils of an E-waste dismantling area in China. Environment International. 2013;**54**:50-58. DOI: 10.1016/j.envint.2013.01.006

[50] Vermeulen K, Covaci A, D'Havé H, Van den Brink NW, Blust R, De Coena W, et al. Accumulation of

background levels of persistent organochlorine and organobromine pollutants through the soil earthworm hedgehog food chain. Environment International. 2010;**36**:721-727. DOI: 10.1016/j.envint.2010.05.006

[51] Coelho C. Transfer and effects of brominated flame retardants (BFRs) on three plant species and one earthworm species in anthroposoils [thesis]. University of Lyon; 2019. p. 285; HAL Id: tel-02420943; https://tel.archives-ouvertes.fr/tel-0242094

[52] Li B, Chen H, Sun H, Lan Z. Distribution, isomerization and enantiomer selectivity of hexabromocyclododecane (HBCD) diastereoisomers in different tissue and subcellular fractions of earthworms. Ecotoxicology and Environmental Safety. 2017;**19**:326-334. DOI: 10.1016/j.ecoenv.2017.01.004

[53] Li B, Zhu H, Sun H, Xu J. Effects of the amendment of biochars and carbon nanotubes on the bioavailability of hexabromocyclododecanes (HBCDs) in soil to ecologically different species of earthworms. Environmental Pollution. 2017;**222**:191-200. DOI: 10.1016/j.envpol.2016.12.057

[54] Chen X, Gu JQ, Wang YF, Gu XY, Zhao XP, Wang XR, et al. Fate and omethylating detoxification of tetrabromobisphenol A (TBBPA) in two earthworms (Metaphire guillelmi and Eisenia fetida). Environmental Pollution. 2017;**227**:526-533. DOI: 10.1016/j.envpol.2017.04.090

[55] Chen X, Gu XY, Zhao XP, Ma X, Pan YN, Wang XR, et al. Species-dependent toxicity, accumulation, and subcellular partitioning of cadmium in combination with tetrabromobisphenol A in earthworms. Chemosphere. 2018;**210**:1042-1050. DOI: 10.1016/j.chemosphere.2018.07.106

[56] Li B, Yao TQ, Sun HW, Zhang YW, Yang JR. Diastereomer-and enantiomer-specific accumulation, depuration, bioisomerization and metabolism of hexabromocyclododecane (HBCDs) in two ecological species of earthworms. Science of the Total Environment. 2016;**542**:427-434. DOI: 10.1016/j.scitotenv.2015.10.100

[57] Morrison DE, Robertson BK, Alexander M. Bioavailability to earthworms of aged DDT, DDE, DDD, and Dieldrin in soil. Environmental Science & Technology. 2000;**34**(4):709-713. DOI: 10.1021/es9909879

[58] Kinney CA, Furlong ET, Kolpin DW, Burkhardt MR, Zaugg SD, Werner SL, et al. Bioaccumulation of pharmaceutical and other anthropogenic waste indicators in earthworms from agricultural soil amended with biosolid or swine manure. Environmental Science & Technology. 2008;**42**:1863-1870. DOI: 10.1021/es702304c

[59] Kinney CA, Campbell BR, Thompson R, Furlong ET, Kolpin DW, Burkhardt MR, et al. Earthworm bioassays and seedling emergence for monitoring toxicity, aging and bioaccumulation of anthropogenic waste indicator compounds in biosolids-amended soil. Science of the Total Environment. 2012;**433**:507-515. DOI: 10.1016/j.scitotenv.2012.06.097

[60] Carter LJ, Garman CD, Ryan J, Dowle A, Bergström E, Thomas-Oates J, et al. Fate and uptake of pharmaceuticals in soil-earthworm systems. Environmental Science & Technology. 2014;**48**:5955-5963. DOI: 10.1021/es500567w

[61] Ma WC, van Kleunen A, Immerzeel J, de Maagd PGJ.

Bioaccumulation of polycyclic aromatic hydrocarbons by earthworms: Assessment of equilibrium partitioning theory in situ studies and water experiments. Environmental Toxicology and Chemistry. 1998;**17**(9):1730-1737. DOI: 10.1897/1551-5028(1998)017<1730:BOPAHB>2.3.CO;2

[62] Ville P, Roch P, Cooper EL, Masson P, Narbonne J-F. PCBs increase molecular related activities (lysozyme, antibacterial, hemolysis, proteases) but inhibit macrophage-related functions (phagocytosis, wound healing) in earthworms. Journal of Invertebrate Pathology. 1995;**65**(3):217-224. DOI: 10.1006/jipa.1995.1033

[63] Bu YQ, Luo YM, Shan ZJ, Teng Y, Li ZG. Effect of dioxin-like PCBs on physiological activities of earthworms (Eisenia fetida). Zhongguo Huanjing Kexue/China Environmental Science. 2010;**30**:699-704

[64] Henriksson S, Bjurlid F, Rotander A, Engwall M, Lindström G, Westberg H, et al. Uptake and bioaccumulation of PCDD/Fs in earthworms after in situ and in vitro exposure to soil from a contaminated sawmill site. Science of the Total Environment. 2017;**580**:564-571. DOI: 10.1016/j.scitotenv.2016.11.213

[65] Law K, Halldorson T, Danell R, Stern G, Gewurtz S, Alaee M, et al. Bioaccumulation and trophic transfer of some brominated flame retardants in a Lake Winnipeg (Canada) food web. Environmental Toxicology and Chemistry. 2006;**25**:2177-2186. DOI: 10.1897/05-500r.1

[66] Law RJ, Allchin CR, De Boer J, Covaci A, Herzke D, Lepom P, et al. Levels and trends of brominated flame retardants in the European environment. Chemosphere. 2006;**64**:187-208. DOI: 10.1016/j.chemosphere.2005.12.007

[67] Braekevelt E, Tittlemier SA, Tomy GT. Direct measurement of octanol–water partition coefficients of some environmentally relevant brominated diphenyl ether congeners. Chemosphere. 2003;**5**(7):563-567. DOI: 10.1016/S0045-6535(02)00841-X

[68] Zhang S, Xu X, Wu Y, Ge J, Li W, Huo X. Polybrominated diphenyl ethers in residential and agricultural soils from an electronic waste polluted region in South China: Distribution, compositional profile, and sources. Chemosphere. 2014;**102**:55-60. DOI: 10.1016/j.chemosphere.2013.12.020

[69] Currier HA, Fremlin KM, Elliot JE, Drouillard KG. Bioaccumulation and biomagnification of PBDEs in a terrestrial food chain at an urban landfill. Chemosphere. 2020;**238**:124577. DOI: 10.1016/j.chemosphere.2019.124577

[70] Matscheko N, Tysklind M, de Wit C, Bergek S, Andersson R, Sellström U. Application of sewage sludge to arable land-soil concentrations of polybrominated diphenyl ethers and polychorinated dibenzo-p-dioxins, dibenzofurans, and biphenyls, and their accumulation in earthworms. Environmental Toxicology and Chemistry. 2002;**21**(12):2515-2525. DOI: 10.1897/1551-5028(2002)021<2515:AOSSTA>2.0.CO;2

[71] Sellström U, De Wit CA, Lundgren L, Tysklind M. Effect of sewage-sludge application on concentrations of higher-brominated diphenyl ethers in soils and earthworms. Environmental Science & Technology. 2005;**39**:9064-9070. DOI: 10.1021/es051190m

[72] Marvin GH, Tomy GT, Armitage JM, Arnot JA, McCarty L, Covaci A, et al. Hexabromocyclododecane: Current understanding of chemistry,

environmental fate and toxicology and implications for global management. Environmental Science & Technology. 2011;**45**:8613-8623. DOI: 10.1021/es201548c

[73] Janák K, Covaci A, Voorspoels S, Becher G. Hexabromocyclododecane in marine species from the Western Scheldt estuary: Diastereoisomer- and enantiomer-specific accumulation. Environmental Science & Technology. 2005;**39**(7): 1987-1994. DOI: 10.1021/es0484909

[74] Hong HZ, Shen R, Liu WX, Li DM, Huang LM, Shi DL. Developmental toxicity of three hexabromocyclododecane diastereoisomers in embryos of the marine medaka Oryzias melastigma. Marine Pollution Bulletin. 2015;**101**: 110-118. DOI: 10.1016/j.marpolbul.2015.11.009

[75] de Bruyn AMH, Gobas FAPC. The sorptive capacity of animal protein. Environmental Toxicology and Chemistry. 2007;**26**(9):1803-1808. DOI: 10.1897/07-016R.1

[76] Mäenpää K, Leppänen MT, Figueiredo K, Mayer P, Gilbert D, Jahnke A, et al. Fate of polychlorinated biphenyls in a contaminated lake ecosystem: Combining equilibrium passive sampling of sediment and water with total concentration measurements of biota. Environmental Toxicology and Chemistry. 2015;**34**(11):2463-2474. DOI: 10.1002/etc.3099

[77] Nakamura M, Yoshikawa H, Tamada M, Fujii Y, Kaneko N, Masunaga S. Bioaccumulation of PCDD/Fs and dioxin-like PCBs in the soil food web of fallow rice fields in Japan. Organohalogen Compounds. 2007;**69**: 1452-1455. Available from: http://risk.kan.ynu.ac.jp/publish/nakamura/nakamura200709.pdf

[78] Reinecke AJ, Nash RG. Toxicity of 2,3,7,8-TCDD and short-term bioaccumulation by earthworms (oligochaeta). Soil Biology and Biochemistry. 1984;**16**(1):45-49. DOI: 10.1006/eesa.1997.1608

[79] OECD (Organization for Economic Co-operation and Development). Test No. 207: Guideline for the Testing of Chemicals: Earthworms, Acute Toxicity. Paris (France): OECD Publishing; 1984. p. 9. DOI: 10.1787/20745761

[80] ISO (International Organization for Standardization). Draft ISO-17512: Soil Quality and Avoidance Test for Evaluating the Quality of Soils and the Toxicity of Chemicals. Test with Earthworms (Eisenia fetida/andrei). Geneve, Switzerland: OECD; 2005

[81] OECD (Organization for Economic Co-operation and Development). Test No. 222: Earthworm Reproduction Test (Eisenia fetida/Eisenia andrei), OECD Guidelines for the Testing of Chemicals 2004. p. 18. DOI: 10.1787/9789264070325-en

[82] Hardy ML, Aufderheide J, Krueger HO, Mathews ME, Porch JR, Schaefer EC, et al. Terrestrial toxicity evaluation of decabromodiphenylethane on organisms from three trophic levels. Ecotoxicology and Environmental Safety. 2011;**74**:703-710. DOI: 10.1016/j.ecoenv.2010.10.027

[83] EC (European Commission). European Union Risk Assessment Report: Bis(pentabromophenyl ether). 1st Priority List. Vol. 17. EUR 20402 EN. European Chemicals Bureau, Institute of Health and Consumer Protection, European Commission; 2002. Available from: https://echa.europa.eu/documents/10162/da9bc4c4-8e5b-4562-964c-5b4cf59d2432

[84] Shi S, Zhang L, Yang W, Zhou L, Dong L, Huang Y. Levels and spatial distribution of polybrominated diphenyl ethers (PBDEs) in surface soil from the Yangtze River Delta, China. Bulletin of Environmental Contamination and Toxicology. 2014;**93**:752-757. DOI: 10.1007/s00128-014-1365-8

[85] Xie X, Wu Y, Zhu M, Zhang YK, Wang X. Hydroxyl radical generation and oxidative stress in earthworms (Eisenia fetida) exposed to decabromodiphenyl ether (BDE-209). Ecotoxicology. 2011;**64**(1):993-999. DOI: 10.1007/s10646-011-0645-x

[86] Zhang W, Liu K, Chen L, Chen L, Lin K, Fu R. A multi-biomarker risk assessment of the impact of brominated flame retardant-decabromodiphenyl ether (BDE209) on the antioxidant system of earthworm Eisenia fetida. Environmental Toxicology and Pharmacology. 2014;**38**(1):297-304. DOI: 10.1016/j.etap.2014.06.007

[87] Liang R, Chen J, Shi Y, Lu Y, Sarvajayakesavalu S, Xu X, et al. Toxicological effects on earthworms (Eisenia fetida) exposed to sublethal concentrations of BDE-47 and BDE-209 from a metabolic point. Environmental Pollution. 2018;**240**:653-660. DOI: 10.1016/j.envpol.2018.04.145

[88] Zhang W, Liu K, Li J, Chen L, Lin K. Uptake and depuration kinetics of lead (Pb) and biomarker responses in the earthworm Eisenia fetida after simultaneous exposure to decabromodiphenyl ether (BDE209). Ecotoxicology and Environmental Safety. 2015;**113**:45-51. DOI: 10.1016/j.ecoenv.2014.11.014

[89] Dendooven L, Alvarez-Bernal D. Contreras-Ramos SMEarthworms, a means to accelerate removal of hydrocarbons (PAHs) from soil? A mini-review. Pedobiologia. 2011;**54**:5187-5192. DOI: 10.1016/j.pedobi.2011.08.006

[90] Liu K, Chen L, Zhang W, Lin K, Zhao L. EPR detection of hydroxyl radical generation and oxidative perturbations in lead-exposed earthworms (Eisenia fetida) in the presence of decabromodiphenyl ether. Ecotoxicology. 2015;**24**:301-308. DOI: 10.1007/s10646-014-1378-4

[91] Zhao Y, Sun L, Li Q, Yan X, Li Z, Liu B. Use of integrated biomarker response for evaluating antioxidant stress and DNA damage of erathworms (Eisenia fetida) in decabromodiphenyl ethane-contaminated soil. Environmental Pollution. 2020;**264**: 114706. DOI: 10.1016/j.envpol.2020.114706

[92] Bruns E, Egeler P, Roembke J, Scheffczyk A, Spoerlein P. Bioaccumulation of lindane and hexachlorobenzene by the oligochaete Enchytraeus albidus (Enchytraeidae, Oligochaeta, Annelida). Hydrobiologia. 2001;**463**(1):185-196. DOI: 10.1023/A:101315981006

[93] Belfroid A, Vandenberg M, Seinen W, Hermens J, Vangestel K. Uptake, bioavailability and elimination of hydrophobic compounds in earthworms (Eisenia andrei) in field-contaminated soil. Environmental Toxicology and Chemistry. 1995;**14**: 605-612. DOI: 10.1002/etc.5620140408

[94] Jager T, Van der Wal L, Fleuren RHLJ, Barendregt A, Hermens JLM. Bioaccumulation of organic chemicals in contaminated soils: Evaluation of bioassays with earthworms. Environmental Science & Technology. 2005;**39**:293-298. DOI: 10.1021/es035317o

Chapter 5

Ecological Risk Index, Implications, and its Bioconcentration Ratio in Soils Contaminated with Heavy Metals

Anahí C. Meza García and José G. Chan Quijano

Abstract

Soil contamination by heavy metals is a problem that is increasing worldwide due to human activities, and it is important to bear in mind that its toxicity and abundance in ecosystems can be serious. The remediation of these contaminated environments should be a priority in all countries, since the consequences that have been recorded are harmful to human and environmental health. In humans, it causes different diseases and even metals bioaccumulate in the human body, as well as in plants and animals, which come as food to the tables of many homes. Therefore, the objective of this chapter is to show the implications and consequences that heavy metals cause. In addition, show evaluation methods and techniques, such as the ecological risk index, to generate data that help investigations and possible alternatives for sustainable remediation. The impact that this theme must be relevant in the penta helix; because all the actors (government, society, business, academia, and civil associations) must cooperate and get involved for inclusive sustainability that provides solutions for the recovery of contaminated ecosystems, as well as for the care of human and environmental health.

Keywords: toxicology, bioaccumulation, acid soils, socio-environmental, remediation

1. Introduction

The impact of heavy metal pollution by urban growth, agricultural, mining, and oil activities has generated environmental problems in soil and aquatic systems, affecting biotic and abiotic components [1, 2]. In addition, anthropogenic activities increase the mobilization and distribution of heavy metals in ecosystems, if the standards allowed by international regulations are exceeded, they result in a serious problem for the health of living beings and the natural environment [3, 4].

Heavy metals are characterized by their high density, which is greater than 5 g ml^{-1} [5–7]. In the natural environment, heavy metals exist in a geogenic form, that is, they come from the parent rock, from volcanic activity, or from the leaching of mineralization [8]. Under normal conditions, most of the potentially toxic metals are found fixed by geological considerations and in very insoluble chemical forms, therefore, they do not represent a danger for the biota, but they become a toxic danger when there is bad management of these (anthropogenic activities) [1].

The introduction of these elements in large quantities to nature causes the breakdown of the balance of the ecosystem, causing the contamination of soils and aquifers; in aquifers, they reach the food web [1–3]. In the food web, heavy metals remain suspended in the water column and are deposited in the sediments [9, 10], which are absorbed by the benthic species, these, in turn, serve as food for the species of the pelagic system, for which create bioaccumulation routes between aquatic organisms [11–13].

The bioaccumulation of heavy metals in the soil is determined by internal factors (chemical characteristics of the element and compounds they form) and by external factors (edaphic processes and environmental conditions), as well as their dispersion in the soil (**Table 1**) [1, 14].

A: Acrisols, B: Cambisols, C: Podzoluvisols, F: Ferralsols, K: Castanozems, L: Luvisols, G: Greyzems, N: Nitosols, P: Podzols, Q: Arenosols, R: Regosols, S: Solonetz, V: Vertisols, W: Planosols, X: Xerosols, Y: Yermosols, Z: Solonchaks. Source: Taken and modified from [1, 14–16].

Processes	Description	Soil unit	Accumulation of heavy metals	Heavy metal dispersion
Podzolization	This process encompasses the leaching of Al and Fe, together with organic matter, from the surface areas and its accumulation in the deep areas of the profile.	R, Q, Y	Co, Cu, Mn, Ni, Ti, V y Zr (illuviation)	B, Ba, Br, Mn, Rb, Cd, Cr, I, Li, Se, Sr, V, Zr
Aluminization	This is a surface treatment, which belongs to the category of diffusion coatings, applied mainly to nickel- and/or cobalt-based superalloy components.	B, L, M, W, A	Co, Mn, Mo, V (gley)	B, Ba, Br, Cu, I, Se, Sr
Sialitization	Process that originates aluminum clays at the expense of silicates.	V, K	B, Ba, Cu, Mn, Se y Sr	—
Laterization	It consists of a process of generalized and deep chemical weathering in which silica and bases are extracted, by leaching, from the parent rock, in which iron and aluminum concretions are produced.	A, F, K	B, Ba, Cu, Mn, Se y Sr	—
Alkalization	It is the excess exchangeable sodium in the soil. As its concentration increases in the soil, it begins to replace other cations.	Z, S, X	B, Co, Cr, Cu, Mo, Ni, Se, Zn, V	—
Hlydromorphy	Edaphic process in which the transformations caused by the presence of a permanent layer of water take place.	—	Ba, B, Co, Cu, I, Mn, Mo, Se, Sr, U	B, Br, Co, Cu, Mn, Ni, U, V

Table 1.
Accumulation and dispersion of some heavy metals according to some edaphic processes.

2. Implications and alternatives

Therefore, heavy metals cause poor soil quality and nutrient loss [17]. Similarly, despite these damages, poor physical soil structure, high concentrations of heavy metals, and nutrient deficiency, contaminated areas often contain unique microbial communities and plant species [18–20]. These act as phytoextractors and bioaccumulators of heavy metals; these characteristics will depend on the number of plant species in the affected area and the biomass of the plants [20, 21].

For example, the acute toxicity in *Agave tequilana* plants against metallic ions, such as Cu^{2+}, Cd^{2+}, and Co^{2+}. Among their results, they show that these metals bioaccumulated in the leaves of the seedlings, so these succulent plants have a remarkable tolerance to high concentrations of the different ionic metals tested, including both micronutrients and toxic metals, as well as the ability to transport these metals in high quantities to air tissue [22].

In aquatic plants, such as *Typha latifolia*, a study was carried out in search of removing heavy metals, such as Cu, Ni, and Zn. The number of metals removed by plants in the sediment was considerably lower (1%), mainly due to the small development of biomass. However, the contribution of these plants is that they can bioaccumulate pollutants in the roots, stems and leaves [23].

In crop plants, for example, with *Hordeum vulgare,* Cd was found, which may be causing toxic effects on the nucleolus of plant cells and affects the expression of nucleolar proteins, nucleolin, and fibrillarin [24]. In *Triticum aestivum*, high levels of Ni (29.3 $\mu g\ g^{-1}$) and Cd (0.22 $\mu g\ g^{-1}$) were found, which puts food safety at risk [25] due to the amounts of contaminants that may be present, bioaccumulating in plants and crop fruits.

Portulaca grandiflora was examined as a possible vegetation for green roofs. It was found that this plant has a hyperaccumulating potential of heavy metals, such as Al, Cd, Cr, Cu, Fe, Ni, Pb, and Zn, within its tissues 40 days after being exposed to the contaminated area, and it also has the ability of metal translocation (Al, Cu, Fe, and Zn) [26].

On the other hand, some metals, such as Co, Cu, Fe, Mn, Mo, Ni, and Zn, are essential micronutrients for plants but become toxic at high concentrations. Its role as an essential micronutrient may be as a cofactor or activator of specific enzymes or to stabilize organic molecules [7, 27].

Other heavy metals, such as Cd, Pb, Cr, Hg, Ag, U, and Au, are nonessential for plant operation. However, they are metals that cause damage to plant species and are bioaccumulated in roots, stems, leaves, fruits, and seeds [7, 28–32].

When there is a regeneration of plant species—mostly grasses—in the areas impacted by heavy metals, these plants can be induced or introduced by human actions [33]. In addition, there is a mutualistic relationship between soil particles and plant roots [34], creating a biotrophic interaction with the ecosystem affected by heavy metals, increasing the ability of plants to absorb these pollutants according to the physiological performance of the plants [35, 36].

3. Phytoremediation and ecological risk index

There are plants that could hyperaccumulate heavy metals from contaminated soils. During this bioaccumulation process, plants enter a state of stress under the different metals, for example, with Cd, the responses of hyperaccumulating plants differ in morphological responses and physiological processes, such as photosynthesis and respiration, uptake, transport, and assimilation of minerals, and nitrogen, and absorption, and transport of water, which contributes to its ability to accumulate, and detoxify Cd [37]. In addition, it affects human health since it is quickly absorbed by plants and can affect crops.

The potential ecological risk index is the process of examining a hazard to avoid or reduce it, that is, it will focus on adverse events that affect environmental health caused by toxic chemical substances [38].

The three aspects that make up the triad of this potential ecological risk in heavy metal contamination are chemical characterization (concentration of contaminants), toxicity (genetic and histopathological damage, enzymatic stress, and mortality), and ecological integration (richness and population distribution of species, **Figure 1**) [7, 27, 39–41].

This triad of potential ecological risk in heavy metal contamination goes hand in hand with the basic strategies for plants to establish themselves in contaminated soils (**Table 2**) [1, 42].

The absorption that plants have is mainly through the roots; this absorption can be both passive (non-metabolic) and active (metabolic). Passive absorption occurs by the diffusion of ions from an external solution into the root endodermis. Instead, active absorption is carried out with metabolic energy down a chemical gradient; at the concentrations present in soil solutions, uptake of heavy metals by plant roots is controlled by metabolic processes [1, 43].

However, in the fauna, the damage is even more severe, since various diseases such as hepatitis, respiratory problems, cancer, and neurological disorders, among others, affect many parts of the body due to the different levels of toxicity, which also affects the cellular level [12, 13, 44]. Studies have been carried out at the level of muscle tissue and liver, mostly in fish and other aquatic species, finding medium and high toxic levels [13, 45–47], which means that these products are ingested as food (without knowing

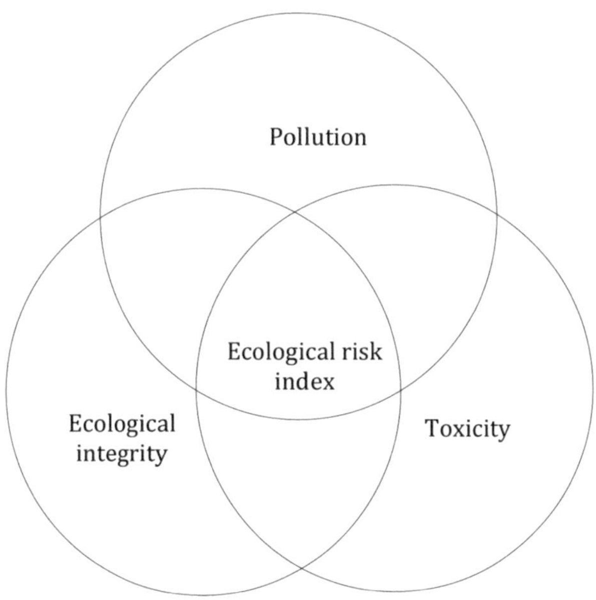

Figure 1.
Descriptive scheme of the triad of integrity to evaluate soils contaminated with heavy metals. Source: own elaboration based on [38, 39].

Strategy	Characteristic
Exclusion	It consists in those plants that must avoid the accumulation of metals in their aerial parts, preferentially accumulating them in the root system.
Accumulation	It consists of the ability of plants to accumulate metals in their tissues proportionally to the concentrations present in the environment.
Hyperaccumulation	It consists of the ability of the plant to accumulate high concentrations of metals in its tissues (stems and leaves).

Source: own elaboration based on [42].

Table 2.
Strategies of multiple tolerance of terrestrial and aquatic plants for their establishment in soils contaminated with heavy metals.

Heavy metal	Toxicokinetic	Distribution and storage in the body	Health effects	Bibliographic reference
Pb	It enters the human body through respiratory, oral, and skin routes.	After absorption, lead passes through the blood and is distributed throughout the body; a dynamic exchange is then established between the different tissues and organs.	It affects the cardiovascular system and damages the central and peripheral nervous system. In adults, 94% of the lead load is found in the skeleton, in children it is 73%. It can cause encephalopathy, memory loss, poor attention, irritability, headache, muscle tremors, clumsiness, hallucinations, fatigue, ataxia, seizures, and coma, among others.	[43]
Hg	Absorption is through respiratory, oral, and skin routes.	It is distributed in red blood cells and plasma in approximately equal proportions. It accumulates mainly in the kidney, as well as the gastrointestinal tract, the squamous epithelium of the hair and skin, and some glands, such as the thyroid, pancreas, sweat glands, testicles, and prostate.	Acute intoxication occurs causing tremors, paresthesia, memory loss, hyperexcitability, erethism, and reduction in reflexes. Also, it causes bronchitis, interstitial pneumonitis, pulmonary edema, metallic taste, increased salivation, stomatitis, gingivitis, cough, chest pain, diarrhea, vomiting, and hemorrhage.	[49]
Cd	It has entry through the respiratory and oral routes.	Once absorbed, it passes into the blood and is transported to all organs and tissues, mainly bound to hemoglobin inside erythrocytes.	Genetically affects the stages of development and functional status of the organism and affects the efficiency of nutrients. It has negative effects on the kidneys (renal efficiency), among others.	[50]

Heavy metal	Toxicokinetic	Distribution and storage in the body	Health effects	Bibliographic reference
Cr	It enters the body by inhalation, ingestion, and to a lesser extent through the skin.	It reaches the intestine and passes into the blood, where it is distributed to the different organs.	The effects depend on the valence state of Cr exposure (Cr III or Cr VI), Cr VI crosses the cell membrane, causing conjunctivitis, tearing, and pain. Cr III damages the skin, causing ulcers and redness. They affect the respiratory system, the gastrointestinal tract, and cause cancer, among other symptoms.	[51, 52]
As	The main entry of As into the body is through the gastrointestinal tract, the respiratory tract, and through the skin is low.	It accumulates mainly in the liver, kidney, lung, and arm. They are also deposited by keratin in hair and nails.	Acute exposure alters the gastrointestinal, cardiovascular, nervous, renal, and hepatic systems. Also, it is possible to observe vasodilation, hyperemia, and edema due to capillary damage and a drop in blood pressure. It causes skin lesions, alterations in the cardiovascular system, alterations in the metabolism of the heme group, neurological effects, effects on the immune system, and on the endocrine system, it causes cancer and genotoxicity.	[53]
Be	The main route of absorption is the respiratory one, by the other routes it is difficult, in the cutaneous absorption it is insignificant.	It circulates through the blood (colloidal phosphate), a part of the absorbed dose is incorporated into the skeleton and in the lungs.	It has effects on the respiratory and skin systems. Causes chronic beryllium disease (it is an irreversible, long-lasting disease of progressive severity; it damages the liver, spleen, and heart).	[54]

Table 3.
Toxic effects in humans due to the ingestion of food contaminated with some heavy metals.

what they contain pollutants), it would be participating in a bioaccumulation network in the human body, which would be presenting in acute or chronic diseases.

The bioaccumulation of heavy metals, such as Cd, Cr, Ni, Pb, V, and Zn, in suspended loads, sediments, primary producers, mollusks, crustaceans, and fish can sometimes be moderate, but even so they pose an ecological risk to the ecosystem [2]. Within the bioaccumulation of heavy metals, these can be ingested by consuming contaminated food and at the same time follow this accumulation process in humans [48]. In humans, the intake of food contaminated by heavy metals estimates serious health effects (**Table 3**).

Therefore, it is necessary to influence government policies and programs that promote the safety of human and environmental health within the gradients of food consumption from orchards, milpas, and production fields near the contaminated areas and due to the complexity of the risk that is run by the movements of pollutants in the soil and the atmosphere. Similarly, it is important to evaluate the perception of health benefits and risks due to the influence of contaminants [55].

4. Conclusions

Heavy metals are highly harmful to human and environmental health depending on their toxicity and abundance.

Ecosystems continue to be contaminated with heavy metals, which should be a red point and national security in several countries for the search for sustainable remediation.

The ecological risk index is a method that helps to evaluate the problem of heavy metal contamination, it is also an alternative to identify the level of contamination in the ecosystem or in human health.

The absorption and bioaccumulation of pollutants in plant, animal and human tissues cause diseases from the mildest to the most serious that can cause death.

Acknowledgements

We thank Dr. Manuel J. Cach Pérez, Dr. Mirna Valdéz Hernández, Dr. Eliane Ceccon, Dr. Manuel Mendoza Carranza, and Dr. Hans Van der Wall for their contributions and observations on the paper. To the Autonomous University of Guadalajara (UAG) for the support given and financial management to carry out the publication.

Conflict of interest

The authors declare no conflict of interest.

Author details

Anahí C. Meza García and José G. Chan Quijano*
Departamento de Ciencia y Tecnología, Universidad Autónoma de Guadalajara, Villahermosa, Tabasco, Mexico

*Address all correspondence to: jose.chan@edu.uag.mx

References

[1] Bautista-Zúñiga F. Introducción al estudio de la contaminación del suelo por metales pesados. Yucatán, México: Ediciones de la Universidad Autónoma de Yucatán; 1999. p. 57

[2] Mendoza-Carranza M, Sepúlveda-Lozada A, Dias-Ferreira C, Geissen V. Distribution, and bioconcentration of heavy metals in a tropical aquatic food web: a case study of a tropical estuarine lagoon in SE Mexico. Environmental Pollution. 2016;**210**:155-165. DOI: 10.1016/j.envpol.2015.12.014

[3] Jahnke A, MacLeod M, Wickström H, Mayer P. Equilibrium sampling to determine the thermodynamic potential for bioaccumulation of persistent organic pollutants from sediment. Environmental Science and Technology. 2014;**48**(19):11352-11359. DOI: 10.1021/es503336w

[4] Rodríguez-Martínez HL, Peña-Manjarrez M, Gutiérrez-Reyes AV, González-Trevizo CL, Montes-Fonseca SL, López-Avalos GG. Biorremediación de arsénico mediada por microorganismos genéticamente modificados. Terra Latinoamericana. 2017;**35**:353-361

[5] Duffus JH. "Heavy metals" – a meaningless term? (IUPAC Technical Report). International Union of Pure and Applied Chemistry. 2002;**74**(5):793-807. DOI: 10.1351/pac200274050793

[6] Brusseau MI, Famisan GB, Artiola JF. Chemical contaminants. In: Artiola JF, Pepper IL, Brusseau M, editors. Environmental Monitoring and Characterization. California, U.S.A: Elsevier, Inc.; 2004. pp. 299-312. DOI: 10.1016/B978-0-12-064477-3.X5000-0

[7] Lambers H, Chapin FS III, Pons TL. Plant Physiological Ecology. 2nd ed. Springer Science+Business Media, LLC: New York, USA; 2008. p. 605. DOI: 10.1007/978-0-387-78341-3

[8] Galán-Huertos E, Romero-Baena A. Contaminación de suelos por metales pesados. Macia. 2008;**10**:48-60

[9] Hernández-García Y, Sosa-Cerecedo M, Moreno M, Alcalá J, Puga S. Evaluación de la contaminación por metales pesados y arsénico en sedimento en embalses del estado de Chihuahua, México. Revista Latinoamericana de Recursos Naturales. 2008;**4**(2):89-94

[10] Aguilar-Ucán CA, Montalvo-Romero C, Ramirez-Elías MA, González-Barrera C. Metales pesados en sedimentos del arroyo "La Caleta" de Ciudad del Carmen, Campeche, México: estudio preliminar. Revista Latinoamericana de Recursos Naturales. 2009;**5**(3):232-237

[11] Páez-Osuna F, Frías-Espericueta MG. Bioacumulación, distribución y efectos de los metales pesados en los peneidos. In: Páez-Osuna F, editor. Camaronicultura y medio ambiente. México: Universidad Nacional Autónoma de México, El Colegio de Sinaloa; 2001. pp. 245-269

[12] Páez-Osuna F. Fuentes de metales en la zona costera marina. In: Botello AV, Rendón-von Osten J, Gold-Bouchot G, Agraz-Hernández C, editors. Golfo de México contaminación e impacto ambiental: diagnóstico y tendencias. 2nd ed. México: Universidad Autónoma de Campeche, Universidad Nacional Autónoma de México, Instituto Nacional de Ecología; 2005. pp. 329-342

[13] Páez-Osuna F. Efectos de los metales. In: Botello AV, Rendón-von Osten J, Gold-Bouchot G, Agraz-Hernández C, editors. Golfo de México contaminación e impacto ambiental: diagnóstico y tendencias. 2nd ed. México: Universidad Autónoma de Campeche, Universidad Nacional Autónoma de México, Instituto Nacional de Ecología; 2005. pp. 343-359

[14] Kabata-Pendias A, Pendias H. Trace elements in soils and plants. 3rd ed. Florida, USA: CRC Press LLC; 2001. p. 432. DOI: 10.1201/9781420039900

[15] Dorronsoro C. Micromorfología de suelos. Segunda parte: programa interactivo multimedia para el autoaprendizaje de la interpretación de preparaciones microscópicas de suelos. In: Universidad de Granada, editor. Información acerca del suelo. España: Departamento de Edafología y Química Agrícola, Universidad de Granada; 2016. p. 3.

[16] FAO. Propiedades del suelo: portal de suelos de la FAO [Internet]. 2017. Available from: https://goo.gl/jHvrTX [Accessed: 15 September 2017]

[17] García C, Moreno JL, Hernández MT, Polo A. Metales pesados y sus implicaciones en la calidad del suelo. In: Valladares F, editor. Ciencia y Medio Ambiente. España: Centro de Ciencias Medioambientales; 2002. pp. 125-138

[18] Medina VF, Maestri E, Marmiroli M, Dietz AC, McCutcheon SC. Plant tolerances to contaminants. In: McCutcheon SC, Schnoor JL, editors. Phytoremediation: Transformation and Control of Contaminants. Canada: Wiley-Interscience, Inc.; 2003. pp. 189-232

[19] Barrutia O, Artetxe U, Hernández A, Olano JM, García-Plazaola JI, Garbisu C, et al. Native plant communities in an abandoned Pb-Zn mining area of northern Spain: implications for phytoremediation and germplasm preservation. International Journal of Phytoremediation. 2011;**13**(3):256-270. DOI: 10.1080/15226511003753946

[20] Burges A, Epelde L, Blanco F, Becerril JM, Garbisu C. Ecosystem services and plant physiological status during endophyte-assisted phytoremediation of metal contaminated soil. Science of the Total Environment. 2017;**584-585**:329-338. DOI: 10.1016/j.scitotenv.2016.12.146

[21] Barrutia O, Epelde L, García-Plazaola JI, Garbisu C, Becerril JM. Phytoextraction potential of two *Rumex acetosa* L. accessions collected from metalliferous and nonmetalliferous sites: effect of fertilization. Chemosphere. 2009;**74**:259-264. DOI: 10.1016/j.chemosphere.2008.09.036

[22] Cen-Cen ER, Gómez-Merino F, Martínez-Hernández A. Tolerancia de *Agave tequilana* a altas concentraciones de cationes metálicos divalentes. Polibotánica. 2015;**40**:163-182

[23] Manions T, Stentiford EI, Millner P. Removal of heavy metals from a metaliferous water solution by *Typha latifolia* plants and sewage sludge compost. Chemosphere. 2003;**53**:487-494. DOI: 10.1016/S0045-6535(03)00537-X

[24] Shi Q, Wang J, Jiang Z, Wu H, Wang J, Jiang W, et al. Effects of cadmium on nucleolus in roots tips of *Hordeum vulgare*. Botanical Sciences. 2017;**95**(1):93-102. DOI: 10.17129/botsci.694

[25] Wang C, Li W, Yang Z, Chen Y, Shao W, Ji J. An invisible soil acidification: critical role of soil carbonate and its impact on heavy metal bioavailability. Scientific Reports. 2015;**5**:12735. DOI: 10.1038/srep12735

[26] Vijayaraghavan K, Arockiaraj J, Kamala-Kannan S. *Portulaca grandiflora* as green roof vegetation: plant growth and phytoremediation experiments. International Journal of Phytoremediation. 2017;**19**(6):537-544. DOI: 10.1080/15226514.2016.1267699

[27] Seshadri B, Bolan NS, Naidu R. Rhizosphere-induced heavy metal(loid) transformation in relation to bioavailability and remediation. Journal of Soil Science and Plant Nutrition. 2015;**15**(2):524-548. DOI: 10.4067/S0718-95162015005000043

[28] Prieto-García F, Filardo-Kerstup S, Pérez-Cruz E, Beltrán-Hernández R, Román-Gutiérrez A, Méndez-Marzo M. Caracterización física y química de semillas de Opuntias (*Opuntia* spp.) cultivadas en el Estado de Hidalgo, México. Biagro. 2006;**18**(3):163-169

[29] Prieto-García F, Méndez-Marzo MA, Martínez-Pezina FH, Prieto-Méndez J. Presencia de metales pesados en cultivos del Valle del Mezquital, México. Revista Latinoamericana de Recursos Naturales. 2007;**3**(2):100-110

[30] Rodríguez-Serrano M, Martínez-de la Casa N, Romero-Puertas MC, del Río LA, Sandalio LM. Toxicidad del cadmio en plantas. Ecosistemas. 2008;**17**(3):139-146

[31] Bravo M, Lengua R, Aguirre R, Ale N, Tomas G, Muñante J, et al. Determinación de metales tóxicos en algunos alimentos vegetales de la zona del Valle de Ilo. Revista Peruana de Química e Ingeniería Química. 2014;**7**(1):30-35

[32] Santiago-Montejo PA. Determinación de plaguicidas en muestras de suelo y productos alimenticios de huertos familiares en la zona Ríos de Tabasco [thesis]. Tabasco, México: El Colegio de la Frontera Sur; 2016

[33] Pignataro AG, Levy-Tacher SI, Aguirre-Rivera JR, Nahed-Toral J, González-Espinosa M, González-Arzac A, et al. Natural regeneration of tree species in pastures on peasant land in Chiapas, Mexico. Agriculture, Ecosystems and Environment. 2017;**249**:137-143. DOI: 10.1016/j.agee.2017.08.020

[34] Novel PS. Physicochemical and environmental plant physiology. 4th ed. Oxford, England: Academic Press; 2009. p. 582. DOI: 10.1016/B978-0-12-374143-1.X0001-4

[35] Li Y, Li Q, Chen B. Organic pollutant penetration through fruit polyester skin: a modified three-compartment diffusion model. Scientific Report. 2016;**6**:23554. DOI: 10.1038/srep23554

[36] Jajoo A. Effects of environmental pollutants polycyclic aromatic hydrocarbons (PAH) on photosynthetic processes. In: Hou HJM, Najafpour MM, Moore GF, Allakhverdiev SI, editors. Photosynthesis: Structures, Mechanisms, and Applications. Cham, Switzerland: Springer; 2017. pp. 249-259. DOI: 10.1007/978-3-319-48873-8

[37] Shanying HE, Xiaoe Y, Zhenli HE, Virupax CB. Morphological and physiological responses of plants to cadmium toxicity: a review. Pedosphere. 2017;**27**(3):421-438. DOI: 10.1016/S1002-0160(17)60339-4

[38] Ize-Lema I. La evaluación de riesgo por sustancias tóxicas. Gaceta Ecológica. 2003;**69**:45-56

[39] Long E, Chapman PM. A sediment quality triad: measures of sediment contamination, toxicity and infaunal community composition in Puget Sound. Marine Pollution Bulletin. 1985;**16**:405-415. DOI: 10.1016/0025-326X(85)90290-5

[40] Osman KT. Soils Principles, Properties, and Management. Dordrecht: Springer Science+Business Media; 2013. p. 271. DOI: 10.1007/978-94-007-5663-2

[41] Bonnial E, Sarmiento AG, Nieto JM, Delvalls TA. Caracterización del riesgo ecológico asociado a metales en sedimentos de la cuenca del Río Odiel a través de índices teóricos. Revista de la Sociedad Española de Mineralogía. 2015;**20**:27-28

[42] González-Mendoza D. Mecanismos de tolerancia a elementos potencialmente tóxicos en plantas. In: Alarcón A, Ferrera-Cerrato R, editors. Biorremediación de suelos y aguas contaminadas con compuestos orgánicos e inorgánicos. México: Editorial Trillas; 2013. pp. 159-177

[43] Saldívar-Osorio L, Tovar-Tovar A, Namihira-Guerrero D. Plomo. In: Albert LA, editor. Toxicología Ambiental. Chihuahua, México: Universidad Autónoma de Ciudad Juárez; 2004. pp. 217-239

[44] Palavicino I, Meléndez P, Bergqvist E, Gálvez X. Determinación de metales pesados (Cu, Mo, Zn, Cd, Pb) en hígados de bovinos procedentes de la Región Metropolitana y VI, VII, VIII, IX, X y XI Regiones de Chile. Avances en Ciencias Veterinarias. 1991;**6**(1):7-9. DOI: 10.5354/acv.v6i1.4627

[45] Durá VB, Jaramillo AM, Monerris MM, García AP. Estudio de acumulación de metales pesados en peces, moluscos y crustáceos de consumo humano, del litoral de la provincia de Castellón. España: Fundación Dávalos-Fletcher; 2004. p. 130

[46] Corrales D. Estudio del contenido de metales pesados en dos especies de peces de la zona costera de Montevideo, Uruguay [thesis]. Uruguay: Universidad de la República de Uruguay; 2013

[47] Corrales-Venegas M. Acumulación de metales pesados en bivalvos y sus efectos tóxicos en la salud humana: perspectivas para el estudio en Costa Rica. Revista Pensamiento Actual. 2015;**15**(25):173-181

[48] Lorenzo-Márquez H, Torres-Dosal A, Barba-MacíasE,Ilizaliturri-HernándezCA, Martínez-Salinas RI, Morales-López JJ, et al. Estimación de riesgo de exposición a metales pesados por consumo de plecos (*Pterygoplichthys* spp.) en infantes de comunidades ribereñas de los ríos Grijalva y Usumacinta, México. Revista Internacional de Contaminación Ambiental. 2016;**32**(2):153-164. DOI: 10.20937/RICA.2016.32.02.02

[49] Gutiérrez-Ruiz MC, Bucio-Ortiz L, Souza-Arroyo V. Mercurio. In: Albert LA, editor. Toxicología ambiental. Chihuahua, México: Universidad Autónoma de Ciudad Juárez; 2004. pp. 241-258

[50] Saldívar-Osorio L, Tovar-Tovar A, van der Goes TF. Cadmio. In: Albert LA, editor. Toxicología ambiental. Chihuahua, México: Universidad Autónoma de Ciudad Juárez; 2004. pp. 259-275

[51] Besoain E. Mineralogía de arcillas de suelos. San José, Costa Rica: Instituto Interamericano de Cooperación para la Agricultura; 1985. p. 1216

[52] Albert LA. Cromo. In: Albert LA, editor. Toxicología ambiental. Chihuahua, México: Universidad Autónoma de Ciudad Juárez; 2004. pp. 277-294

[53] Del Razo LM, Quintanilla-Vega B, Albores A. Arsénico. In: Albert LA, editor. Toxicología ambiental. Chihuahua, México: Universidad Autónoma de Ciudad Juárez; 2004. pp. 295-308

[54] Albert LA. Berilio. In: Albert LA, editor. Toxicología ambiental. Chihuahua, México: Universidad Autónoma de Ciudad Juárez; 2004. pp. 309-323

[55] Bernardino-Hernández HU, Mariaca-Méndez R, Nazar-Beutelspacher A, Álvarez-Solís JD, Torres-Dosal A, Herrera-Portugal C. Percepciones del uso de plaguicidas entre productores de tres sistemas agrícolas en Los Altos de Chiapas, México. Acta Universitaria. 2017;**27**(4):26-41. DOI: 10.15174/au.2017.1188

Chapter 6

Effect of the Degree of Soil Contamination with Cd, Zn, Cu and Zn on Its Content in the Forder Crops and Mobility in the Soil Profile

Dorota Pikuła

Abstract

In the micro plot experiment, the effect of varying light and medium soil contamination with heavy metals on their content in rye green fodder, serradella forage and barley, and the migration of Cd, Pb, Cu, and Zn in the soil profile was evaluated. Plants accumulated more heavy metals on light soil and soil with low organic matter content. Under the influence of increasing heavy metal contamination of soils, cadmium content increased more in fodder plants than lead content. After 14 years from the introduction of different rates of metals into the top layer (0–30 cm) of the two soils studied, a relatively large movement of heavy metals in the soil profile occurred. The amount of leached metals depends mainly on the rate of a given element. The more contaminated the soil, the heavier the metals that leached to lower genetic levels of soil. An increase in mobility was obtained alongside an increase in soil contamination with the heavy metals studied. By analyzing the mobility coefficients, they can be ranked in the following decreasing sequence: on light soils: Cd > Cu > Zn > Pb and on medium soils: Cd > Zn > Pb > Cu.

Keywords: fodder plants, pollution level, heavy metals, organic carbon, bioaccumulation index

1. Introduction

One of the major problems of our time is heavy metal contamination of soils associated with the progressive development of civilization [1]. At the same time, more and more emphasis is being placed on the quality of crops for consumption, fodder, and processing. Therefore, from the point of view of modern agriculture, the yield of fodder crops must be correlated with their quality. The content of heavy metals in soils is of great interest to ecologists, biologists, agricultural producers, and environmentalists alike [2]. This issue becomes particularly important under conditions of heavy metal contamination of cultivated soils [3]. In the absence of practical ways to permanently remove excess heavy metals from soils, efforts are made to bind them to the soil in such a way that the possibility of uptake by plants is reduced to a minimum [4]. One way of counteracting the negative effects of heavy metal contamination of soils and plants is the application of organic matter in the form of organic fertilizers of various types, for example, manure, peat, and composts [5]. On the other hand, in the case of moderate contamination, it is proposed to introduce agrotechnical measures to reduce the bioavailability of metals from the soil and their content in plants to a level that does not endanger animal and human health [6] (soil liming).

The natural environment (soils, waters, and plants) is usually contaminated with not one, but several heavy metals. Therefore, the Department of Agricultural Chemistry of the Warsaw University of Life Sciences has undertaken research to assess the impact of the degree of soil contamination with presumably contaminating elements in content in the forder crops. An additional objective of these studies is to assess the impact of various factors on the mobility of individual metals in the soil by determining the mobility coefficients in the conditions of crop management techniques, balanced mineral and organic fertilization, and soil liming. For this purpose, in 1987, a unique microplot experiment was established on soils artificially contaminated with copper, zinc, lead, and cadmium oxides (up to the pollution level of classes I, II, and III, according to the regulation of the Minister of the Environment) [7]. The soils were diversified in terms of pH (through liming), organic matter content (through the addition of brown coal), and the grain size composition of the humus level (Ap) (strong clay sand and light silt clay).

A research facility was used for the study, where the soils were contaminated simultaneously with several heavy metals, that is, Cu, Zn, Pb, and Cd, with contamination of each of these metals occurring in classes 0, I, II, and III. The soils were characterized by different pH, humus, and fluffy content.

The paper assesses the influence of artificial contamination of soils with heavy metals of light and medium soils on the movement of heavy metals deep into the soil profile, depending on selected physical and chemical properties and content thereof in fodder plants.

Four metals were selected for the study, that is, Cu, Zn, Pb, and Cd included in the group of metals with a very high risk for the environment [8] and forage. In terms of threat to feed quality, Cd and Pb are the most dangerous, while the toxic ones include Cu and Zn [9].

2. Heavy metal content of soils and plants

2.1 Sources and status of heavy metal contamination of soils in Poland

Heavy metals are generally referred to those metals, which possess a specific density of more than 5 g/cm^3 and adversely affect the environment and living organisms [10]. These metals are quintessential in maintaining various biochemical and physiological functions in living organisms when in very low concentrations; however, they become noxious when they exceed certain threshold concentrations. Zinc and copper are elements essential for life and development of organisms, while cadmium and lead have no specific physiological function and are even harmful to plant development and human health [11]. Under conditions unaltered by man, the natural content of heavy metals in soils depends primarily on the abundance of the bedrock, on weathering and soil-forming processes, and on the granulometric composition [12].

The natural source of heavy metals in soils is the parent rock from which the soils were formed. In Poland, 80% of arable land has a natural and 17.6% elevated heavy metal content. Consequently, only 3% of Poland's arable land is described as contaminated [13, 14]. Weakly and moderately contaminated soils account for 2.7%, while higher and very highly contaminated soils occupy 0.27% and 0.08% of the agricultural area, respectively [15]. Agricultural pollution results from the inappropriate use of soil conditioners produced from sewage sludge. Therefore, the agricultural use of municipal wastewater and industrial sludge for fertilization and liming of soils must be under constant control of the quantity and quality of harmful components introduced. Soil contamination can be localized or diffused [16]. Soil contamination of a local nature affects small areas, and their occurrence is most often associated with intensive industrial activities, inadequate waste storage and treatment, and the extraction and processing of minerals (e.g., metal ores) [11]. Threats from pollution of a point source nature mainly concern industrial and post-industrial, urbanized areas and areas near transport routes and landfills [17].

2.2 Heavy metal toxicity to plants

The toxicity of heavy metals is due to their biochemical role in metabolic processes and the extent to which they are absorbed and excreted by living organisms [18, 19]. The detrimental effect of heavy metals on living organisms is to induce enzyme dysfunction. Heavy metals can block enzyme systems and induce physiological changes that lead to cell and tissue death. In plants, damage to the root system occurs, followed by stunted growth. Depending on the sensitivity of the plants to heavy metals, chlorosis and necrosis appear on the leaves, which are not always characteristic of individual heavy metals. Plants readily take up heavy metals from the soil. An excess of trace elements in the plant causes metabolic disturbances. This usually reduces yield quality, plant germination, and delays emergence. Excessive amount of heavy metals reduces the intensity of photosynthesis, inhibits plant growth, and, above all, causes underdevelopment and distortion of roots [20]. Low concentrations of toxic elements do not adversely affect plant yield. However, at excessive levels, heavy metals have a toxic effect, accumulating in the tissues, retarding plant growth and development, resulting in the plant yielding less or dying completely. Heavy metals damage plants, deform and contribute to weakening the plants' resistance to diseases and pests. The assimilative surface area is reduced due to numerous necrotic spots as a result of the harmful effects of heavy metals, which consequently reduces the efficiency of photosynthesis [21]. When the tolerance limit of heavy metal content in fodder and consumer plants is exceeded, symptoms of poisoning can be expected in animals and humans [22, 23].

The phytotoxicity of heavy metals is largely genetically determined, which is why there are strong species and cultivar variations among plants [24]. This is related, among other things, to the unequal action of resistance mechanisms in individual species and even varieties. One such mechanism in plants tolerant to excess lead is a reduction in the movement of lead from the roots to the aboveground parts. Lead in such plants is stored in the cell walls of the roots as hardly soluble inorganic compounds [25, 26]. Another important mechanism concerning cadmium and copper involves the binding of these elements by a specific plant protein, phytochelatin [27]. It is generally believed that dicotyledonous plants are more susceptible to the harmful effects of heavy metals compared to monocotyledonous plants [28]. Considering the way, the plants are used, it is found that vegetables are less resistant to the toxic effects of heavy metals, compared to agricultural plants [29, 30].

Heavy metals have the ability to move through the trophic chain and, once the tolerance limit is exceeded in the case of fodder or food crops, increased incidence of chronic or acute disease in animals and humans is to be expected. In the environment, heavy metals are present in plants in trace amounts of less than 0.1% d.m. Despite such small amounts, their excessive accumulation can lead to contamination of the entire ecosystem. Kabata-Pendias [28, 31] distinguished four groups of trace elements in terms of their potential threat to the environment. Thus, cadmium, lead, zinc, and copper are categorized as being of very high risk, while nickel belongs to the group of medium environmental risk [32]. In terms of threat to feed quality, Cd and Pb are the most hazardous, while Cu and Zn are among the toxic ones [9].

2.3 Forms and mobility of heavy metals in soil

Heavy metals enter the soil mainly as insoluble compounds [33]. They then undergo different transformations in the soil depending on their physical and chemical properties. The behavior of heavy metals is governed by two opposing groups of phenomena. The first involves processes that reduce the ability of the metals in question to migrate. These obviously include sorption processes, precipitation of insoluble compounds from solution, and biogenic accumulation. This results in the accumulation of heavy metals in the topsoil horizons. The second group is the processes that increase the mobility of heavy metals: desorption, solubility, and decomposition of organic compounds [34, 35].

The solubility of heavy metals in the soil solution, on the one hand, and accumulation in the soil, on the other hand, are mainly influenced by soil properties such as pH, granulometric composition, sorption complex capacity, organic matter content and type, and oxidation-reduction conditions [35].

As the pH value increases, the ability of the soil to bind metal ions into forms that are difficult to dissolve increases. At low pH, all heavy metals, except molybdenum, are available to plants. Therefore, their phytoavailability poses great problems in acid and light soils, which constitute more than 60% of Polish soils [36]. On the other hand, according to Czarnowska and Szymanowska-Sieńczewska [37], the amount of heavy metals present in soluble forms depends primarily on their total content in soils. According to Stępień et al. [38, 39], the cadmium content of soils is significantly influenced by the organic matter content. These authors argue that organic matter currently acts as a buffer in nature for pollutants of anthropogenic origin. The buffering capacity of soil toward these pollutants is mainly based on their immobilization in insoluble compounds and their complete or partial detoxification.

The strong sorption properties of the soil, especially of the organic matter it contains, meaning that heavy metals are largely retained in the topsoil humus layer. The danger that a metal poses to the environment depends not only on the amount in which it is introduced into the soil, but above all, on the chemical form in which it occurs [40].

Heavy metals are found in soils in different forms, which determine their solubility and bioavailability [41] state that, for agricultural purposes, five fractions in which heavy metals occur in soils can be separated:

Exchangeable fraction: These are metals that are weakly bound to the soil solids and readily released into the soil solution.
Carbonate fraction: It includes metals bound to carbonates, readily passing into soil solution at low soil pH.
Iron and manganese oxide fraction: It forms the binder or coating, covering the mineral parts of the soil. Metals in this group are sensitive to changes in redox potential.
Organic fraction: Metals bound to soil organic matter and released gradually during mineralization, which, when spread out over time, results in the periodic retention of heavy metals in forms unavailable to plants.
Another fraction is the *residue fraction*: These are metals associated with silicate minerals and are found in stable compounds.

The authors consider the exchangeable and carbonate fraction to be easily soluble and available to plants. Heavy metals in water-soluble and exchangeable form should be considered as easily mobile in the soil available to plants, while the others are taken up to a lesser extent. Carbonates under suitable conditions (low pH) and oxides (low redox potential) can supplement the soluble and exchangeable form [42]. Under conditions favoring oxidation, metals bound to organic matter can also become available to plants. The active form remains in a state of very rapidly establishing equilibrium with the mobile form and, in most chemical tests, all of the active forms and some parts of the mobile form are determined.

Considering ecological and agronomic aspects, attention should be paid mainly to reactive (readily soluble, highly active, and directly available to plants) or potentially mobile forms of heavy metals (potentially available to plants, active, and leached), which are extracted from the soil by means of weakly inert salts or with chelating solutions [43].

The solubility of cadmium in soil solution, compared to other metals, is high [21, 24]. Its mobility is largely related to the source of soil contamination. Thus, in soils formed from cadmium-rich rocks, the mobility of this element is lower than in soils contaminated with it due to industrial activities [45, 46]. In light soils with elevated cadmium content, the proportion of its exchangeable forms may exceed 50% of the total content [47]. Gorlach and Gambuś [48] emphasize that the solubility of cadmium in soil solution and the limitation of its uptake by plants is strongly influenced by soil organic matter. In contrast, according to Ristovic [49, 50], the activation of soluble forms of cadmium is mainly related to the decrease in soil pH. Under conditions of higher pH, the solubility of chemical combinations of cadmium decreases and the adsorption of this element on soil colloids increases [51]. In soils with a pH of 4.5–5.5, cadmium is highly mobile, and at higher values, it becomes immobilized, forming carbonates [52]. Under the influence of the introduction of municipal waste into soils, cadmium becomes mobilized.

Lead, unlike cadmium, is not very mobile in soil and only a small part of it is soluble in mild solvents [53]. This element is rarely present in soil in the form of Pb cation^{+2}, but it forms complex ions $PbOH^+$ and $Pb(OH)_4^{2-}$, which significantly regulate sorption and desorption processes. It is strongly bound by most soil components, especially by Fe and Mn concretions. It is also sorbed by clay minerals, aluminum and iron hydroxides and by organic matter [28]. Weber [54] emphasizes that, in addition to soil type and organic matter content, the solubility of lead is also influenced by soil pH. Increased calcium content promotes higher soil pH and reduces the uptake of lead by plants. Lead is poorly mobile in contaminated soils, especially light soils, but the proportion of soluble and phyto-available forms is high [55]. The significant amount of this metal in the exchangeable fraction indicates the danger of lead desorption and mobilization from the soil. In acidic soils, on the other hand, lead forms predominate in organic compounds, which can both increase and decrease its migration. Higher solubility of lead occurs in light acidic soil than in less acidic clay soil. The mobile forms of lead in acidic soils occur mainly as Pb cations^{2+}, $PbHCO_3^+$, and organic complexes. In alkaline soils, on the other hand, the predominant forms are: $PbOH^+$ and $Pb(CO_3)_2^{2-}$.

Zinc in the soil environment forms complex ions, for example, $ZnOH^+$, ZnO_2^-. All zinc compounds dissolve under the influence of weathering processes (easily soluble form most accessible to plants), at low soil pH, while the released ions form mineral combinations with high mobility. Organic matter sorbs zinc under certain acidic conditions in the soil; at pH 5.8, zinc is bound by 60% humic acids; at pH < 5.8, sorption disappears. The resulting bonds are persistent and, as a result, a fairly significant amount of zinc is observed in the surface horizons of organic and mineral soils. Zinc is one of the most mobile trace elements [56]. This is influenced by both its exchangeable forms and its compounds with organic matter. Organic matter forms fairly stable bonds with zinc, and therefore, there is an accumulation of zinc in the surface horizons of mineral soils and in organic soils. The addition of municipal waste or various zinc compounds to the soil increases the mobile forms of this element. An increase in soil pH decreases its desorption (solubility) in the soil. However, the formation of complex anions and organic-mineral combinations can maintain its high mobility also in alkaline soils. Zinc concentrations in natural soil solutions range from 60 to 2200 g/l, with the highest values in light acidic soils.

Copper in the soil is not very mobile, so its loss by leaching is generally low at approx. 30 g, year/ha. In soils, copper is strongly bound by clay minerals and organic matter, and is also lost as sulfides, sulfates, carbonates, giving poorly mobile forms [57]. Therefore, only a small proportion of copper occurs in soils in mobile forms—in soluble and exchangeable forms. Organic matter plays the main role in binding copper, but microorganisms also play an important role in binding this element. The bioavailability of copper bound to an organic substance depends on the fines weight. Thus, organic compounds with a low weight, which are released during the decomposition of organic matter or are introduced together with municipal waste increase the mobility of copper. In contrast, peat organic matter, as well as humic acids, immobilize copper, more or less permanently.

The content of soluble forms of copper in the topsoil of our soils is 1–5 ppm, but varies from 1–30 ppm, representing a few to several percent of its total content [58]. The bioavailable copper content depends on the amount and type of organic matter. There is less plant-available copper in organic soils because it is strongly bound there. All soil minerals show a high copper binding capacity, which is dependent on the environmental pH.

In an acidic environment there is a high mobility of cationic forms of copper—Cu^{+2}, $CuOH^+$, $Cu_2(OH)_2^{2+}$, and as pH increases, mobile anionic forms prevail—$Cu(OH_3)_2^{2-}$, $Cu(OH)_{3^-}$, $Cu(OH)_4^{2-}$ [51]. Copper is subject to constant accumulation in the surface horizons, is strongly bound there and is not subject to displacement deep into the profile. The phytoprecipitation of copper can be reduced with lime and organic matter (peat).

2.4 Factors affecting heavy metal content in plants

Industrial pollution causes the introduction of many heavy metals into soils, which combine with the sorption complex or pass into the soil solution. Dynamic equilibrium states are established in the sorption complex, changing under the influence of external factors, that is, mineral fertilization, precipitation, change in pH [59]. In turn, the uptake of heavy metals by plants from soils is a function of many factors, both biological (plant properties) and environmental (soil properties, climatic factors), [28]. On the other hand, the different biochemical properties of elements determine their susceptibility to bioaccumulation (uptake by plants). Thus, under similar soil conditions, cadmium and zinc are more intensively taken up by plants than manganese, lead or copper. The factors considered to have the most significant influence on the plant uptake of heavy metals are the total content of potentially bioavailable elements in the soil, the concentration of the elements concerned in the soil solution and their relative quantitative proportions, and the flow of elements from the solid to the liquid phase of the soil and then to the roots. As a result of the above-mentioned factors, nickel and lead are considered to have low mobility and therefore low bioavailability, while cadmium and zinc are considered to have high mobility. Through biological selection processes of chemical elements, plants can regulate, to a certain extent, their chemical composition. However, the action of physiological barriers is largely limited with respect to heavy metals and is also subject to breakdown under the influence of their excessive concentration [60]. Therefore, the content of heavy metals in terrestrial plants is generally positively correlated with their occurrence in soils, as well as in lower-lying bedrock, which is used for geochemical prospecting studies.

Under natural conditions, there is a great variation among plant species and varieties in the intensity and selectivity of trace element uptake. Species differences are large even among vegetation occurring in the same habitat. Metal concentrations also vary according to the specific plant part and developmental stage. Plants take up heavy metals, not only through the root system from the soil, but also through the leaves and other above-ground parts, on which compounds of these elements fall in the form of dissolved rainwater or dust. An example of this is lead in cereal plants (in the above-ground parts and even the grain), which is approximately 90% derived from atmospheric pollution fallout. Therefore, determining the source of excessive trace elements in plants as well as in the trophic chain is not easy and generally requires detailed research [28].

Plants are an important link in the movement of metals from the soil to animal and human organisms. For concentrations of metals tolerated by plants but harmful to humans and animals, for example, cadmium, zinc, and lead, the action of the biological barrier of selective uptake of elements by plants is not effective [61]. Trace element mobilization processes in soils are of particular importance in cases of soil contamination by these elements. In general, elements of anthropogenic origin are found in more active forms, and the degree of their immobilization depends on the properties of the soils, especially the reaction and granulometric composition [51]. General regularities are known that determine the degree of harmfulness of heavy metal contamination, depending on soil properties. There is always a significantly higher bioavailability of metals from very light acid soils compared to their bioavailability from heavy soils with neutral pH [62]. Calcium carbonate might act as a strong absorbent for heavy metals and could be complex [63]. The effect of soil type on metal mobility is similarly influenced, and therefore light acidic soils, which have very low buffering and filtering properties, have an increased risk of groundwater and groundwater contamination. Heavy soils, on the other hand, sometimes referred to as "resistant" soils to chemical contamination, can accumulate and immobilize significant amounts of certain elements. Metal fixation in heavy soils is intermittent, as any change in the conditions of the soil environment, for example, increase in acidity and changes in the quality or quantity of organic matter and in the redox potential, can affect their mobilization and therefore easy bioavailability [64].

Various factors influence the bioavailability of heavy metals in plants, but the most important of these is the quantitative content of these elements (hence the establishment of limit numbers to determine the amount that is safe for humans, animals, and the environment) and the physicochemical properties of the soil, which determine the form in which the metals in question are present and therefore their mobility. Research has shown that the three most important soil properties in this respect are the granulometric composition, pH, and humus content of the soil. These properties, together with the metal content of the soil, determine which contamination class the soil belongs to. The granulometric composition of the soil (content of clayey parts) not only determines the content and mobility of trace elements but also shapes the sorption capacity of the soil (restriction of mobility of macro and microelements) [15]. The heavy metal binding capacity of aluminum, iron, and manganese oxides. The heavy metal binding capacity of the above-mentioned oxides is influenced by: their quantitative and qualitative content and the soil pH. An acidic soil pH reduces the sorption capacity of aluminum, iron, and manganese oxides. Lead combines most easily with the oxides in question, while manganese oxides have the highest sorption capacity [47]. Also, the presence of clay minerals in the soil limits the availability of heavy metals to plants. As the humus and clay mineral content increases, the bioavailability and solubility of heavy metals decrease. This relationship is particularly evident in acidic soils. This is particularly true for lead, which is most highly bound to clay minerals. This is followed by copper, zinc, and cadmium [65]. A large amount of plant-available zinc can be immobilized by introducing sufficient montmorillonite into the soil [66]. The same clay mineral can be used for cadmium, lead, and copper. In addition to montmorillonite, soils can be treated for heavy metals using zeolites and bentonites, both synthetic and mineral [67].

An important factor limiting the availability of toxic elements is the soil pH. Heavy metals dissolve more readily in acidic soil and are therefore better available to plants. However, they differ in this respect [62]. The best example is zinc and cadmium, for which a drop in pH < 6.0 results in increased activity, which is not observed to such a great extent for lead and copper. Therefore, liming is an essential treatment, especially for reducing the availability of zinc and cadmium to plants. Czuba et al. [68] report, "the effect of pH on the uptake of heavy metals depends on the plant species and the metal." This statement was confirmed in an experiment in which the effect of pH on the heavy metal content of aboveground parts of sunflower and maize was studied. The most significant differences were seen when examining the phytoavailability of zinc: in sunflower, (with a decrease in pH < 6.5, the zinc content in the above-ground parts increased significantly: at pH = 3.0, it was about 800–900 mg/kg d.m., at pH = 6.5, it was about 80–90 mg/kg d.m.), in maize, the difference was not so significant, (zinc content in the above-ground parts, at pH = 3.0, was about 150–200 mg/kg d.m.; and at pH = 6.5, about 70–80 mg/kg d.m.). Similar differences were observed for other metals, that is, lead, copper, cadmium, and nickel [68].

The natural occurrence of humus compounds in the soil and organic compounds, that is, humus or compost, also reduce the degree of heavy metal contamination of plants. For this reason, among others, a common cultivation procedure used by farmers is to fertilize the soil with peat, compost, or manure. The better the organic matter is decomposed, the greater the effectiveness of the treatments. This is because humus binds heavy metals into stable complexes [69].

3. Materials and methods

3.1 Study area location

The experiment was carried out on microplots at the experimental station in Skierniewice (Poland) (Lat:51°58′ N; Long:20°10′ E), which belongs to the Faculty of Agriculture and Biology at the Warsaw University of Life Sciences under natural climatic conditions. The climate at the site is temperate with a mean annual rainfall for the years 1921–2020 at 536.1 mm, a mean annual temperature 8.1°C, and insolation at 1714 h. This climate is considered as Dfb, according to the Köppen-Geiger climate classification (warm humid continental climate).

The experimental factors in the experiment were: two types of soil (light soil-loamy sand with 17% of particles <0.02 mm and medium soil-light dusty clay with 25% of particles <0.02 mm), three levels of Corg concentration (6, 9, and 12 g $C{\cdot}kg^{-1}$ of the soil), four levels of heavy metal contamination of soils Zn, Pb, Cd, and Cu (natural content—0, increased content—I, weak contamination—II, medium contamination—III).

The microplots were supplied with vitrified clay pipes with a diameter of 0.4 m and a height of 1.2 m high. They were buried vertically in the soil, which in 1987, were filled with layers of Albic Luvisol soil [70]: (1) ochric (Ap) 0–30 cm (the surface horizon that has been plowed or cultivated)—17% of particles <0.02 mm; (2) luvic (Et) 30–50 cm (the subsurface horizon of eluvial loss of silt and clay)—13% of particles <0.02 mm; (3) agric (Bt) below 50 cm (the subsurface horizon of illuvial accumulation of clay and humic substances)—25% of particles <0.02 mm.

In 2004, the arable soil layer was removed from each pot to a depth of 30 cm. In half of the microplots, a soil material of strong clay sand composition was used, while in the second half of light dusty clay. Appropriate amounts of calcium carbonate, lignite, and heavy metals (Cd, Cu, Zn, and Pb) were added to the 0–30 cm layer. As a result, soils representing four levels of contamination with four abovementioned heavy metals were obtained. Single doses of individual heavy metals per microplot (56 kg of soil) were: zinc—3.2 g in the form of ZnO, lead—28 g in the form of PbO, copper—1.3 g in the form of CuO, and cadmium—35 mg in the form of $CdSO_4$–H_2O. In this way, research objects with different acidification levels (pH 4.2, 4.9, and 5.9) were created, three levels of organic carbon (approx. 6, 9, and 12 $g{\bullet}kg^{-1}$), two levels of floating parts (17% and 25%), and four levels of soil contamination with heavy metals [natural content (0), elevated (I), polluted in the 1st degree (II), polluted in the 2nd degree (III)]. The classification mentioned above enables us to assess the extent of soil contamination with heavy metals taking their bioavailability into account. The experiment consisted of 216 ground microplots. In the experiment, different plant species were cultivated in the following years of the experiment (cereals, maize, rapeseed, mustard seed, and coarse-seeded legumes).

3.2 Collection and processing of soils

Soil samples were taken for testing in 2018 after the harvest of spring barley for grains. Soil samples were collected at the end of the growing season (September–October) from 0 to 25 cm soil layer. Next, the soil was prepared for analysis in accordance with the ISO11465 standard. The soil material was determined for: organic C by direct method PN-EN 15936, pH in the suspension KCl ($mol{\cdot}d.m.^{-3}$) PN-EN 10390, heavy metals were determined in three levels (top layer Ap, Et, Bt) in the extract of 1 M HCl, while in level Ap 3 with methods: total (in royal water acc. to PN-ISO 11466) and in 1 mol·d.m.$^{-3}$ HCl [71]. The content of cations in the solution was determined by the AAS method. In order to validate the method for accuracy and precision, certified reference material was used.

The movement of heavy metals in the soil profile was determined on the basis of the mobility coefficient CM (increase in heavy metal in the Bt and Et layer compared to the control). The obtained results were analyzed statistically by means of variance analysis (using Tukey's test) and single-factor regression. Statistical calculations were performed using the program statgraphics plus). Results are the means from three replications.

3.3 Plant material analyses

- Total heavy metal content of Zn, Pb, Cd, and Cu by atomic absorption spectrometry. The plant samples were combusted in a mixture of acids HNO_3, $HClO_4$, and H_2SO_4 in a volume ratio of 20:5:1 using the microwave technique in a closed system.

- The following indices were calculated for the forage crops studied [72]:

- The tolerance index (Ti), which indicates the degree of growth inhibition of the test plant when grown on contaminated substrate according to the formula:

- Ti = plant d.m. of plants on contaminated substrate/plant d.m. of plants on control substrate.

- A contamination index (C), which indicates the amount of contamination of a plant on a relative scale, that is, the ratio of the toxic ion in a plant growing on contaminated soil to its level in control plants growing under uncontaminated environmental conditions according to the formula:

  C = metal content of contaminated plant/metal content of control plant.

- The metal bioaccumulation index (A), which determines the increase in heavy metal taken up by the plant relative to its increase in the substrate according to the formula:

  A = increase in metal content of plant/increase in metal content of substrate.

4. Results

4.1 Influence of soil contamination levels of heavy metals on their content in forage crops

4.1.1 The content of Zn, Pb, Cd, and Cu in rye forage

On both soils, an increase in the level of heavy metal contamination of the soils resulted in an increase in the heavy metal content of the rye green fodder. The heavy metal content in rye green fodder at pollution level III was many times higher compared to sites with natural content. The degree of effect of soil contamination with heavy metals on their content in rye green fodder varied for individual metals and depended significantly on the soil species. On the light soil, as the contamination increased, the content of Cd and Pb increased the most in rye green fodder and the content of Cu increased the least. A similar relationship occurred on medium soil.

Critical heavy metal contents in feed: Cd—1, Pb—10, Cu—30, and Zn—100.

The series of changes in the increase of heavy metal content in rye forage on both light and medium soil at pollution level III takes the following form: Cd > Pb > Zn > Cu. It is noteworthy that already at level I of soil contamination with heavy metals, that is, at relatively low contamination, a significant accumulation of these metals in rye green fodder was found, especially of Cd and Zn, that is, mobile elements (**Table 1**). In rye green fodder, exceedance of the permissible heavy metal contents occurred for Zn on light soil at pollution levels I, II, and III, on medium soil at pollution levels II and III contamination levels, and for Cd on light soil at level III of soil contamination with heavy metals.

The content of Zn, Pb, Cd, and Cu in serradella.

In both light and medium soil, increasing the amount of heavy metals in the soil resulted in an increase in their content in serradella. The degree to which heavy metal contamination of the soil affected the content of heavy metals in serradella varied for individual metals and depended, as in the case of rye, on the soil species. On light soil, serradella increased the most in Zn, Cd, and Pb content, and on medium soil, Cd and Zn content increased the most. In contrast, Cu content increased least in serradella on light and medium soil. Serradella grown on light soil significantly accumulated more heavy metals compared to plants grown on medium soil (**Table 2**).

Species soils	Level pollution. Soils with metals Heavy	Heavy metal content							
		Zn		Pb		Cd		Cu	
		$mg{\cdot}kg^{-1}$ d.m.	%	$mg{\cdot}kg^{-1}$ d.m.	%	$mg{\cdot}kg^{-1}$ d.m.	%	$mg{\cdot}kg^{-1}$ d.m.	%
Light	0	45.9	100	0.30	100	0.03	100	5.2	100
	I	**105.6**	230	0.41	137	0.15	500	7.0	135
	II	**202.7**	442	1.60	533	0.50	1667	15.0	441
	III	**276.5**	602	3.20	1067	**1.10**	3667	14.8	435
	On average	157.7	—	1.40	—	0.44	—	10.5	—
Average	0	34.3	100	0.20	100	0.02	100	3.4	—
	I	72.0	210	0.30	150	0.10	500	4.7	138
	II	**112.2**	327	0.80	400	0.21	1050	8.7	256
	III	**184.6**	538	2.10	1050	0.94	4700	12.9	379
	on average	100.8	—	0.83	—	0.32	—	7.4	—

Table 1.
Heavy metal content of life for green fodder in relation to light and medium soil contamination levels.

Species soils	Level pollution and soils with metals Heavy	Heavy metal content							
		Zn		Pb		Cd		Cu	
		mg·kg^{-1} d.m.	%	mg·kg^{-1} d.m.	%	mg·kg^{-1} d.m.	%	mg·kg^{-1} d.m.	%
Light	0	**180.1**	100	3.42	100	**2.80**	100	8.9	100
	I	**199.3**	111	3.71	108	**3.60**	129	11.3	127
	II	**288.0**	160	4.64	136	**4.00**	143	12.9	145
	III	**344.9**	192	5.44	159	**4.42**	158	13.1	147
	On average	253.1	—	4.30	—	3.70	—	11.5	—
Average	0	**140.9**	100	3.00	100	**1.35**	100	7.7	100
	I	**170.5**	121	3.23	108	**2.34**	173	9.7	126
	II	**270.7**	192	3.90	130	**2.80**	207	11.1	144
	III	**317.3**	225	5.20	173	**3.72**	275	12.0	156
	On average	224.9	—	3.82	—	2.55	—	10.1	—

Table 2.
Heavy metal content of serradella in relation to light and medium soil contamination levels.

Critical heavy metal contents in feed: Cd—1, Pb—10, Cu—30, Zn—100

The series of changes in the heavy metal content of serradella on light soil at pollution level III takes the following form: Zn > Cd > Pb > -Cu, and on medium soil: Cd > Zn > Pb > Cu. In addition, already at level I of soil contamination with heavy metals, slight contamination resulted, as in the case of rye forage, also a significant accumulation of these metals in serradella, especially Cd and Cu. At contamination level I, there was no such large increase in the content of toxic elements, as was the case with rye green fodder, indicating that the increase in heavy metal content in plants depends on the plant species. Exceedance of the permissible heavy metal contents in serradella occurred for Zn and Cd on light and medium soil at all contamination levels.

The content of Zn, Pb, Cd, and Cu in barley

The increase in the content of heavy metals in barley grain that takes place under the influence of an increase in soil contamination coincides in terms of the extent of the changes to serradella and rye greens. On light soil, at soil contamination, level III heavy metals in barley grain increased the most in Cd and Pb content and the least in Cu content. In contrast, on medium soil, Cd and Pb content increased the most, while Zn content was the least (**Table 3**).

Critical heavy metal contents in feed: Cd—1, Pb—10, Cu—30, and Zn—100.

The series of changes in the heavy metal content of barley grain on light soil at pollution level III takes the following form: Cd > Pb > Zn > Cu, and on medium soil the following shape takes place: Cd > Pb > Cu > Zn. In barley straw on light soil at pollution level III, Cd and Zn content increased the most

Species soils	Level pollution. Soils with metals Heavy	Heavy metal content							
		Zn		Pb		Cd		Cu	
		mg·kg^{-1} d.m.	%	mg·kg^{-1} d.m.	%	mg·kg^{-1} d.m.	%	mg·kg^{-1} d.m.	%
Light	0	26.2	100	0.32	100	0.12	100	2.7	100
	I	46.1	176	0.44	137	0.25	208	3.6	133
	II	64.6	247	0.74	231	0.60	500	5.5	204
	III	82.2	314	1.50	469	0.90	750	7.2	267
	On average	**54.8**	—	**0.74**	—	**0.50**	—	**4.7**	—
Average	0	22.1	100	0.24	100	0.10	100	1.90	100
	I	37.2	168	0.33	137	0.15	150	2.74	144
	II	49.0	222	0.54	225	0.43	430	4.02	212
	III	70.8	320	0.85	354	0.70	700	6.20	326
	on average	**44.8**	—	**0.50**	—	**0.34**	—	**3.71**	—

Table 3.
Heavy metal content of barley grain in relation to light and medium soil contamination levels.

and Cu the least, while on medium soil, as in grain, Cd and Pb content increased the most and Cu the least. In barley straw on light soil, at pollution level III, the decreasing series of heavy metal contents takes the following form: Cd > Zn > Pb > Cu, and on medium soil the following shape: Cd > Pb > Zn > Cu (**Table 4**).

Species soils	Level pollution. Soils with metals Heavy	Heavy metal content							
		Zn		Pb		Cd		Cu	
		mg·kg^{-1} d.m.	%	mg·kg^{-1} d.m.	%	mg·kg^{-1} d.m.	%	mg·kg^{-1} d.m.	%
Light	0	30.3	100	0.53	100	0.21	100	3.1	100
	I	72.7	240	0.80	151	0.50	238	4.4	142
	II	99.9	330	1.05	198	0.85	405	6.6	213
	III	**220.0**	726	3.20	604	**1.70**	809	10.2	329
	on average	105.7	—	1.40	—	0.81	—	6.1	—
Average	0	24.6	100	0.31	100	0.13	100	2.5	100
	I	55.8	227	0.60	193	0.35	269	3.4	136
	II	80.1	326	0.81	261	0.60	461	4.9	196
	III	**160.4**	652	2.05	661	1.00	769	7.7	308
	on average	80.2	—	0.94	—	0.51	—	4.6	—

Table 4.
Heavy metal content of barley straw in relation to light and medium soil contamination levels.

Critical heavy metal contents in feed: Cd—1, Pb—10, Cu—30, and Zn—100

In barley grain, there were no exceedance of forage heavy metal limits. However, in barley straw, exceedance occurred for Zn at contamination level III, on light and medium soil, and Cd on light soil.

Average bioaccumulation indices (A).

The average value of the bioaccumulation index (5) of Zn in rye green fodder was 1.12, Cd 0.27, Pb 0.007, and Cu 0.23. The high value of the bioaccumulation index of Zn indicates a high propensity of rye to accumulate this element. In the case of serradella, the highest average bioaccumulation index values were obtained for Cd and Zn, 1.30 and 1.11, respectively, indicating a high predisposition of serradella to accumulate these elements under soil contamination conditions and for Pb and Cu, 0.009 and 0.08, respectively. Significantly lower bioaccumulation index values were obtained for rye grain: Zn 0.36, Cd 0.17, Cu 0.21, and Pb 0.004. This indicates that under soil contamination conditions, grain accumulates lower amounts of heavy metals compared to rye and serradella green fodder. Also, barley straw under conditions of soil contamination with elements accumulates lower amounts of these elements compared to rye and serradella green fodder, as evidenced by lower bioaccumulation coefficients (**Table 5**).

Level of soil pollution	Serradella	Rye green fodder	Barley straw	Barley seeds	Mean
		Zn			
I	1.1	2.2	2.3	1.7	**1.8**
II	1.7	3.8	3.2	2.3	**2.7**
b	2.0	5.7	6.9	3.1	**4.4**
		Pb			
I	1.0	1.4	1.6	1.3	**1.3**
II	1.3	4.8	2.3	2.2	**2.6**
III	1.6	10.9	6.3	4.0	**5.7**
		Cd			
I	1.5	4.8	2.5	1.8	**2.6**
II	1.7	12.5	4.1	4.7	**5.7**
III	2.1	38.6	7.7	7.4	**14.0**
		Cu			
I	1.3	1.3	1.3	1.4	**1.3**
II	1.4	2.6	2.0	2.0	**2.0**
III	1.5	3.2	3.1	3.0	**2.7**

Table 5.
Average bioaccumulation index values (A) of Zn, Cd, Pb, and Cu.

Plant	Zn	Cd	Pb	Cu
Serradella	1.11	1.30	0.009	0.08
Rye green fodder	1.12	0.27	0.007	0.23
Barley grain	0.36	0.17	0.004	0.21
Barley straw	0.74	0.42	0.009	0.07
Mean	**0.83**	**0.54**	**0.007**	**0.15**

Table 6.
Average values of the heavy metal contamination rate (C) of plants.

Average contamination rates (C).

The analysis of the degree of heavy metal contamination of plants shows that, in general, the index is higher on light soils than on medium soils under the same dose of heavy metals. At contamination level I (**Table 6**), the highest contamination degree index values were obtained for Zn in rye green fodder and barley straw, 2.2 and 2.3, respectively, and for Cd in rye green fodder 4.8 and in barley straw 2.5. In contrast, at contamination level III, the highest contamination index values for rye green fodder and barley straw were obtained for Pb and Cd.

4.2 Influence of soil species on heavy metal content of forage crops

Soil type, regardless of the level of contamination, significantly influenced the heavy metal content of forage crops. On medium soil, the heavy metal content in rye forage compared to light soil was 13 to 58% lower. The greatest reduction in content was for Cd by 58% at pollution level II and Pb by 50% at pollution level II, respectively (**Table 7**).

The heavy metal content of serradella on medium soil was 5–52% lower compared to that on light soil. The greatest reduction in content occurred, as in the case of rye forage, for Cd and Pb, by 35% at contamination level I and 16% at contamination level II, respectively (**Table 8**).

Level pollution	Species soils	Heavy metal content ($mg \cdot kg^{-1}$ d.m.**)*								
		Zn		Pb		Cd		Cu		Heavy metal content series
		**	%	**	%	**	%	**	%	
0	Light	45.9	100	0.30	100	0.03	100	5.2	100	Cu > Cd = Pb > Zn
	Average	34.3	75	0.20	67	0.02	67	3.4	65	
I	Light	105.6	100	0.41	100	0.15	100	7.0	100	Cu = Cd > Zn > Pb
	Average	72.0	68	0.30	73	0.10	67	4.7	67	
II	Light	202.7	100	1.60	100	0.50	100	15.0	100	Cd > Pb > Zn > Cu
	Average	112.2	55	0.80	50	0.21	42	8.70	58	
III	Light	276.5	100	3.20	100	1.10	100	14.8	100	Pb > Zn > Cd > Cu
	Average	184.6	67	2.10	66	0.94	85	12.9	87	

**For average pH and Corg.*

Table 7.
Heavy metal content of life for green fodder according to soil species at different levels of contamination.

Level Pollution	Species soils	Heavy metal content ($mg \cdot kg^{-1}$ d.m.**)*								
		Zn		Pb		Cd		Cu		Content series heavy metals
		**	%	**	%	**	%	**	%	
0	Light	180.1	100	3.42	100	2.80	100	8.9	100	Cd > Zn > Cu > Pb
	Average	140.9	78	3.00	88	1.35	48	7.7	86	
I	Light	199.3	100	3.71	100	3.60	100	11.3	100	Cd > Zn > Cu > Pb
	Average	170.5	85	3.23	87	2.34	65	9.7	86	
II	Light	288.0	100	4.64	100	4.00	100	12.9	100	Cd > Pb > Cu > Zn
	Average	270.7	94	3.90	84	2.80	70	11.1	86	
III	Light	344.9	100	5.44	100	4.42	100	13.1	100	Cd > Pb > Zn = Cu
	Average	317.3	92	5.20	95	3.72	84	12.0	92	

**For average pH and Corg.*

Table 8.
Heavy metal content of serradella according to soil species at different levels of contamination.

On the other hand, the average heavy metal content in barley grain was 14–40% lower compared to that on light soil, and the greatest reductions occurred in the amount of Cd, by 40% at pollution level I and Pb by 43% at pollution level III, respectively (**Table 9**).

In barley straw, the average heavy metal content on medium soil compared to light soil was lower by 19–42%, and the greatest reduction in content was for Cd and Pb at pollution level III by 41% and 36%, respectively (**Table 10**).

Conclusion.

Soil species significantly influenced the heavy metal content of the forage crops tested. Serradella, rye forage, and barley contained significantly more heavy metals at a given level of contamination on light soil, compared to the contents found on medium soil. An increase in the content of soil runoff parts significantly reduced the content of all heavy metals tested, with Cd and Pb being the most abundant.

Level pollution	Species soils	Heavy metal content ($mg \cdot kg^{-1}$ d.m.**)*								
		Zn		Pb		Cd		Cu		Content series heavy metals
		**	%	**	%	**	%	**	%	
0	Light	26.2	100	0.32	100	0.12	100	2.7	100	Cu > Pb > Cd > Zn
	Average	22.1	84	0.24	75	0.10	83	1.9	70	
I	Light	46.1	100	0.44	100	0.25	100	3.60	100	Cd > Pb > Cu > Zn
	Average	37.2	81	0.33	75	0.15	60	2.74	76	
II	Light	64.6	100	0.74	100	0.60	100	5.5	100	Cd > Pb = Cu > Zn
	Average	49.0	76	0.54	73	0.43	72	4.0	73	
III	Light	82.2	100	1.50	100	0.90	100	7.2	100	Pb > Cd > Cu = Zn
	Average	70.8	86	0.85	57	0.70	78	6.2	86	

**For average pH and Corg.*

Table 9.
Heavy metal content of barley grain according to soil species at different levels of contamination.

Level of pollution	Soil types	Heavy metal content (mg·kg^{-1} d.m.**)*								
		Zn		Pb		Cd		Cu		Content series heavy metals
		**	%	**	%	**	%	**	%	
0	Light	30.3	100	0.53	100	0.21	100	3.1	100	Pb > Cd > Zn = Cu
	Average	24.6	81	0.31	58	0.13	62	2.5	81	
I	Light	72.7	100	0.80	100	0.50	100	4.4	100	Cd > Pb > Zn = Cu
	Average	55.8	77	0.60	75	0.35	70	3.4	77	
II	Light	99.9	100	1.05	100	0.85	100	6.6	100	Cd > Cu > Pb > Zn
	Average	80.1	80	0.81	77	0.60	71	4.9	74	
III	Light	220.0	100	3.20	100	1.70	100	10.2	100	Cd > Pb > Zn > Cu
	Average	160.4	73	2.05	64	1.00	59	7.7	75	

For average pH and Corg.

Table 10.
Heavy metal content of barley straw according to soil species at different levels of contamination.

4.3 Effect of soil contamination on heavy metal content of light and medium soil profile

4.3.1 Physicochemical characteristics of soils

The actual pH of light soils was 4.0, 4.97, and 5.96, while of medium soils, it was 4.03, 5.08, and 6.05, respectively. The actual Corg. concentration in the light soil was 6.6, 9.2, and 11.9 g–kg^{-1}, while in the medium soils, it was 6.2, 8.9, and 1.8 g–kg^{-1}, respectively.

4.3.2 The relocation of heavy metals in soil profiles

Metal contamination of the soil surface layer suggested an investigation into the migration of heavy metals in soil profiles that was necessary to determine the potential contamination of the lower soil horizons. After 14 years, from the introduction of different rates of metals into the top layer (0–30 cm) of the two soils, a relatively large movement of heavy metals in the soil profile occurred (**Table 11**). The amount of leached metals depended mainly on the rate of a given element. The more contaminated the soil was, the heavier metals leached to lower genetic levels. Contaminated soils always had a higher concentration of individual metals in Et than in Bt level. In the Et layer, the content of the metals tested determined in HCl (1 mol–d.m.$^{-3}$) was comparable to the humus level. Only at the depth below 50 cm (Bt), the content of the studied metal form was much lower than at the surface levels. It indicates that within a few years after the contamination of soils with heavy metals, the largest amounts of them moved to the subsoil layer. However, in the layer below 50 cm, significant amounts of the heavy metals studied were also obtained, which indicates the possibility of their movement into deeper soil layers.

The calculated mobility coefficients of the tested metals determined in 1 M HCl indicate a larger movement of the tested metals in lighter soils than in medium soils (**Table 6**). The highest displacement coefficients were obtained for cadmium, while the lowest was for lead. An increase in mobility was obtained alongside the increase in soil contamination with the heavy metals. By analyses of the mobility coefficients

Soil texture	Level of contamination	Cd		Cu		Zn		Pb	
		Mobility factor CM							
		Et	Bt	Et	Bt	Et	Bt	Et	Bt
Strong loamy sand	0								
	I	2.3^{a}	2.0^{a}	2.1^{a}	1.2^{a}	2.1^{a}	1.4^{a}	1.7^{a}	1.0^{a}
	II	4.6^{b}	5.0^{b}	4.0^{b}	2.7^{b}	4.4^{b}	2.7^{b}	4.0^{b}	2.5^{b}
	III	8.3^{c}	7.0^{c}	9.5^{c}	6.5^{c}	8.3^{c}	7.2^{c}	8.5^{c}	4.8^{c}
	Mean	5.1	4.7	5.2	3.5	4.9	3.8	4.7	2.8
Light silty loam	0								
	I	2.0^{a}	1.3^{a}	1.5^{a}	1.5^{a}	2.2^{a}	1.8^{a}	1.6^{a}	1.2^{a}
	II	4.7^{b}	4.3^{b}	3.2^{b}	2.1^{b}	4.4^{b}	2.3^{b}	3.2^{b}	2.0^{b}
	III	6.8^{c}	5.8^{c}	6.3^{c}	5.2^{c}	6.7^{c}	6.2^{c}	7.8^{c}	4.8^{c}
	Mean	4.5	3.8	3.6	2.9	4.4	3.4	4.2	2.7

a, b, and c—Treatments with the same letter are not significantly different ($p \leq 0.05$)*, Et: the subsurface horizon of eluvial loss of silt and clay, Bt: the subsurface horizon of illuvial accumulation of clay and humic substances.*

Table 11.
The mobility coefficients (CM) for Cd, Cu, Zn, and Pb are determined in HCl (1 mol d.m.$^{-3}$*) in Et and Bt levels in light soil depending on the degree of contamination of the soils with these elements.*

(heavy metal increase in the Bt and Et layers), they can be ranked in the following decreasing sequence: on light soils: Cd > Cu > Zn > Pb and on medium soils: Cd > Zn > Pb > Cu.

5. Discussion

In the study, the heavy metal content was determined only in the parts of the plants used for animal feed. Plants contained the heaviest metals on sites at pollution level III in an acidic environment, on light soil with the lowest organic matter content. Serradella accumulated the most Zn and Cd, which is confirmed by the high values of the bioaccumulation index for these elements, 1.11 and 1.30, respectively. A similar trend occurred in the case of rye green fodder with a Zn bioaccumulation index of 1.12. The value of these indices indicates that the increase in heavy metal concentration in plants may be greater than the content in the soil. Similar results were obtained by Piesak [73, 74] and Curyło and Jasiewicz [75] for vegetables. Also, Chłopecka [76] reports that a 1-unit increment in soil Cd content resulted in a 1.2-unit increase in its content in wheat straw. Literature contains information that the ratio of Cd content in plants to its amount in the soil is between 1 and 10 [44]. In contrast, the lowest values of the bioaccumulation index for Cd and Zn were found for barley grain, 0.36 and 0.17, respectively. In all the plants studied, the lowest values of bioaccumulation indices were obtained for Pb and Cu, 0.007 and 0.15, respectively, indicating that these metals are less well taken up by plants than Cd and Zn. This is confirmed by the research of Alloway and Ayres [44], who report that the value of the bioaccumulation index for Pb does not exceed 0.1. This is also consistent with the thesis that the concentration of heavy metals in the soil depends on the dynamics of the movement of metals from the roots to other plant organs, where accumulation follows the following sequence: grain (seed) < leaf < stem < root, and on the mobility of the metal and its chemical properties. Consequently, lead is immobilized easily in the roots [77]. The metal is retained in cell membranes and easily binds to proteins and fats [78]. According to Horubala [79], the highest concentration of lead in the potato is found in the skin, with a slightly lower concentration in the flesh. The low permeability of Pb from the soil to the plant is related to the strong sorption of this metal in the soil [80]. Wallace and Romney [77] report that zinc accumulates mainly in above-ground parts, as it can move very easily into stems, leaves, and even seeds. A similar trend occurred in our study of serradella and rye greens, with serradella accumulating higher amounts of Zn.

According to Strączyński [81, 82], the distribution and content of copper in plants depend not only on the degree of soil contamination with this element but also on the species and parts of the plant. In his studies, the concentration of copper in generative organs is generally lower (except for cereals and oilseed rape) than the amount accumulated in vegetative parts, and soil contamination with copper significantly differentiated the content of this element in crop plants. As soil quality deteriorates, the copper content in plants increases. Maximum concentrations of this metal in plants were found on sites with severe soil contamination. The highest accumulation is found in hemp leaves, stalks, potato stalks, and flax bags. In contrast, Kucharski et al. [83] report that Cu, at doses of 0 to 80 mg/kg soil, has no toxic effect on faba bean, and the growth of this plant is only inhibited by a dose of 100 mg Cu per kg soil. In our study, Cu content in plants increased with increasing soil contamination, as evidenced by high correlation coefficients (0.70 for serradella, 0.85 for barley straw, 0.68 for barley grain, and 0.77 for rye green fodder), but no exceedances of the permissible Cu content were found in plants. Most likely, significant amounts of it were accumulated by the roots. According to Kabaty-Pendias [28], plant roots retain significant amounts of Cu, both in excess and deficiency. In addition, compared to other elements, it is not very mobile in plants. The low uptake of Cu by plants could also be due to antagonisms between Cu and Zn and Ca. Increased concentrations of these elements in the plant habitat reduce Cu uptake. This is due to their chemical properties because Zn can occupy the same positions as Cu in organic carriers and substitute each other [21]. Also, Chaudhry et al. [84] report that high doses of zinc can cause copper deficiency in plants. The antagonistic effect of Zn with respect to Cu is generally coupled to Cd. Excessively high doses of Cd result in lower uptake of Cu by plants, due to its property of substituting Cu for Cd in enzymes and proteins. In contrast, Traynor and Knezek [85] report that an increase in soil Cd content results in increased uptake of Cu and Fe by plants. On the other hand, studies by Lagerwerff [86] indicate that Cu reduces Cd uptake, while Pb increases its concentration in plants. When considering the Zn and Cd content of the plants studied, it was found that rye greens and serradella accumulated particularly high levels of heavy metals. This is consistent with the hypothesis that plants during their slow growth period, that is, early spring and autumn, accumulate large amounts of cadmium and lead [87, 88]. Also, studies carried out by IUNG showed that the Cd and Pb content in cereals, legumes, and grasses harvested in spring exceeded the threshold levels proposed by IUNG [13, 14] for feed purposes. Also, research carried out in the area of Huta, Częstochowa, showed that buckwheat harvested in spring contained more Zn compared to the content found in buckwheat harvested in autumn. Buckwheat in spring accumulated particularly high amounts of Cd (6.6 mg/kg d.m.) on light and acid soils. Serradella reacted similarly in our study to even low levels of heavy metal contamination in soils. It contained high amounts of Zn and Cd even on metal-free soil, which may be due to its species' predisposition to accumulate heavy metals, as shown, for example, by spinach or lettuce. Similar relationships were shown in a study by Gębski et al. [89], which determined the effect of soil contamination with heavy metals on their content in radish, wheat grain and straw, Chinese cabbage, and potatoes. It was found that the average increase in heavy metals in cabbage as soil contamination increased was greater than in the other plants. In this crop, an increase in soil metal content increases lead and cadmium the most and copper the least. Warda [90] also notes the relationship between plant species and cadmium and nickel content in pasture plants. She reports that timothy growing on black soil accumulates only minimal amounts of cadmium, while cocksfoot, under the same conditions, contains almost four times as much of this element. Also, white clover, which makes up a significant proportion of the pasture sward, accumulates small amounts of cadmium, while common dandelion accumulates more than twice as much cadmium as common cocksfoot and perennial ryegrass growing on organic soil. The increased uptake of cadmium by common dandelion under conditions of low concentrations of this metal in the soil may therefore indicate a danger of its inclusion in the trophic chain as its concentration in the soil increases. Therefore, when selecting plant species for grazing mixtures, the predisposition of individual plant species to uptake heavy metals must be taken into account.

In the literature, statements can be found that cadmium and zinc are more phytotoxic, while lead causes less harm [38, 39, 91]. Our own research did not confirm, unequivocally, which metal was the most toxic, as plants were affected by a total of four heavy metals.

In our study, the Pb content in plants increased with increasing levels of contamination, but, as in the case of Cu, there were no exceedances of its permissible content in forage. Larger amounts of Pb were found in rye and serradella greens and the lowest in barley grain.

Piotrowska and Kabata-Pendias [92, 93] and Kabata-Pendias and Pendias [94] report that lead is a low-mobility element, and under conditions of contamination, it increases mainly in the roots and to a lesser extent in the aboveground parts. A study by Gębski et al. [89] showed that in radish roots, zinc, and lead content increased the most and copper the least. Similarly, it was the case with potatoes, but lead and cadmium content increased the most and Cu content the least. In wheat grain, on the other hand, zinc and copper content increased the most and lead content the least. This is due to the fact that zinc and cadmium are among the mobile elements that accumulate most rapidly in the plant. It is generally believed that dicotyledonous plants are more susceptible to the harmful effects of heavy metals compared to monocotyledonous plants [31, 95]. On the other hand, many authors [29, 30] report that, from the point of view of plant use, vegetables are found to be more sensitive to the toxic effects of heavy metals than agriculturally used plants.

According to some scientists [96], the phytotoxicity of heavy metals is largely genetically determined, which is why there are strong species and cultivar variations among plants. This is related, among other things, to the non-uniform action of resistance mechanisms in individual species and cultivars. In lead-tolerant plants, one such mechanism is to reduce the movement of lead from the roots to the aboveground parts. Lead in such plants is stored in the cell walls of the roots in the form of organic compounds that are difficult to dissolve [25]. In the literature [97], attention is drawn to the antagonism between cadmium and zinc, which is usually expressed as a decreased uptake of cadmium by plants when the amount of zinc in the environment is increased, and a decrease in zinc concentration in plants that take up a lot of cadmium. The phenomenon of parallel uptake of both elements under conditions of high accumulation in the soil is also known. This relationship occurred in our study in the case of serradella, on sites where the soil contained higher amounts of zinc and cadmium, respectively (pollution levels I. and II. and at "0" pollution level, in an acidic environment, with low soil organic matter content). Cadmium concentration in plants, according to many researchers, is a characteristic of species and even their cultivated varieties [97, 98]. Some species can show considerable tolerance to the presence of cadmium in the habitat and are able to take up large amounts of this component. Grasses vary in their ability to take up cadmium. Considering the concentration of this element in dry matter, they can be ranked as follows: Festuca pratensis >Alopecurus pratensis > Festuca arundinacea > Lolium perenne > Dactylis glomerata > Phleum pretense. In all grass species tested, there was an inhibition of growth and development with an increase in cadmium uptake.

When considering the content of tested elements in plants, it should be noted that the content of Zn and Cd in serradella grown on light and medium soil on all combinations including those with natural heavy metal content exceeded the values limiting their use for fodder according to the Regulation of the Minister of Agriculture and Rural Development of 23 January 2007 [99]. Also, too high concentration of Zn in green fodder on light soil at I, II, and III levels of contamination, on medium soil at II and III levels of contamination, and Cd on light soil at III level of contamination disqualifies it for fodder. In barley straw, on the other hand, exceedances of standards occurred only for Zn on combinations at pollution level III on light and medium soil and Cd on light soil at pollution level III. Of the plants tested, only barley grain can be used for animal feed without reservation.

Soil species, regardless of the level of heavy metal contamination, significantly influenced their content in forage crops. On medium soil, the heavy metal content in fodder plants compared to light soil was 13–58% lower in rye forage, 5–52% lower in serradella, 14–40% lower in barley grain and 19–42% lower in barley straw. On medium soil, the Cd and Pb contents decreased the most in plants compared to light soil. This is mainly due to the sorptive properties of medium soil, which immobilizes heavy metals, acting as a barrier preventing plants from being poisoned by them, and it confirms the thesis put forward by Niemyska-Łukaszuk [100] that as the content of floating parts in the soil increases, the mobility of heavy metals decreases, so plants do not take up harmful elements. In soils with a higher content of clay minerals, a significant reduction in the solubility of heavy metals is found, mainly as an effect of their non-specific adsorption [65]. The binding strength of heavy metals by clay minerals decreases in the direction: Pb > Cu > Zn > Cd [65]. Light soils, due to their low content of floating parts, structure, and granulometric composition, have a limited capacity to retain incorporated substances due to their low sorption capacity. In contrast, heavy and medium soils act as a kind of buffer, retaining substances introduced into the soil, including heavy metals. This property allows the medium soil to fix and sorb nutrients and heavy metals. The heavier the soil, the richer it is in clay and silt particles and the greater its sorption capacity, which leads to blocking the uptake of harmful elements by plants [6, 13, 14].

In ore search, the movement of PCE in the soil profile was detected. The highest leaching took place at low soil pH and relatively low organic matter content [62]. In the more contaminated soil, the heavier metals leached deep into the soil profile. The most susceptible to leaching was Cd, while the least was Cu. In the literature, it can be found that more leaching of cadmium and nickel occurs from sand soil than from loam soil [101]. Comparing the results of the research with the data of other authors, it can be stated that higher leaching of heavy metals deep into the soil profile occurs on lighter soils compared to the medium. In Gębski's study [51], heavy metals (Zn and Cd) introduced into the arable layer were easily moved to deeper layers of the soil profile. The mobility of these elements, and thus, their leaching intensity, increased together with the decrease in soil pH. A similar tendency was observed in our own research. However, at high soil pH, the movement of Cd deep into the soil profile is small [102]. In turn, Borowiec et al. report that Cd can accumulate mainly at the accumulation level, although most studies indicate that this element can easily move deep into the soil profile. Many researchers have observed a significant reduction in the leaching of Zn and Cd metals under the influence of an increase in soil pH value and a lack of Pb response to a change in pH [103]. This thesis is confirmed by our own research. There is also no evidence of the movement deep into soil profiles of Zn, Cd, and Pb in soils with a similar pH to the neutral one and having a higher content of floatable parts [104]. The finding that higher movement of Zn and Cd occurs in sandy soils with strong acidity [105] is consistent with the results of our research.

Most literature reports that lead does not leach into the soil profile, but Bowen [105] states that this element does move in the soil, yet very slowly. Kabata-Pendias and Pendias [94], on the other hand, draw attention to the possibility of a serious change in the rate of lead movement in the soil profile due to a decrease in soil sorption capacity or an increase in acidification. Under such conditions, this element can migrate to groundwater, posing a serious toxicological risk. Other researchers from the Czech Republic who determine the mobility of Pb, Zn, and Cd in soils from the Příbram region (Czech Republic) heavily contaminated by metallurgy-two profiles of alluvial soils were closely studied-report that the profile distribution of lead (the least mobile metal) is characterized by a gradual decrease with depth [106].

The literature also states that heavy metals do not leach deep into the soil profile but accumulate in the topsoil layer [107, 108]. In the arable layer of Lublin soils, higher amounts of cadmium are found compared to their subsoil [109]. In their study, a higher Cd concentration was found in the layer of 25–50 than 0–25 cm. This shows that the leaching of heavy metals in natural conditions is lower than in the microplot experiment, where this movement is strong.

6. Conclusions

The heavy metal content of forage plants depended on soil and plant species. The average increase of heavy metals in fodder plants as contamination increased was greater on light soil than on medium soil. As the level of soil contamination increased, their content in forage plants increased. The highest contents of the heavy metals tested were found at soil contamination level III and were many times higher compared to sites with natural contents. Under the influence of low pollution, that is, at level I of soil contamination with heavy metals, the content of Cd and Zn increased the most in fodder plants, and the content of Cu increased the least. In contrast, at pollution level III, Pb content increased in plants. It is noteworthy that already pollution level I, that is, a relatively low contamination of soils with heavy metals, caused a significant increase in the heavy metal content of the tested fodder plants, especially in the case of rye green fodder. The highest average values of heavy metal bioaccumulation rates in plants were obtained for Zn and Cd, indicating a high propensity of plants to assimilate these elements. On the other hand, lower values of bioaccumulation indices for Pb and Cu indicate lower predisposition of plants to uptake of these elements. However, there was no effect of soil species on the value of the heavy metal bioaccumulation index. Individual heavy metals also differed in their indicators of the degree of plant contamination. In terms of this characteristic, they can be ranked in the following order, according to the decreasing value of the index: Cd > Pb > Zn > Cu, which means that under conditions of soil contamination with toxic metals, the relative content of Cd increases the most in plants and Cu the least). Of the plants tested, serradella proved to be the most sensitive to high heavy metal content. Even at "0" contamination level, it accumulated large amounts of Zn and Cd, 180 and 2.8 mg/kg, respectively.

According to the Regulation of the Minister of Agriculture and Rural Development of 23 January 2007 [99] on the permissible content of undesirable substances in feed (Journal of Laws No. 162, item 1704 [110]), the content of Cd in feed should not exceed 1 mg/kg d.m., Zn—100 mg/kg d.m., Pb—10 mg/kg d.m. and Cu—30 mg/kg d.m.

The high Cd and Zn content of serradella and rye green fodder on most combinations, especially on light soil, disqualifies them for use as fodder. Barley grain and straw (with the exception of crops harvested on combinations at contamination level III, as there was an exceedance of the standards in terms of Zn and Cd content there) can be used for animal feed.

The leaching of the studied metals depended on the degree of contamination and soil species. The more contaminated the soil was, the more PCE determined in 1 M HCl leached to lower genetic levels, more to Et than Bt. The highest displacement factors were obtained for cadmium, and the lowest were for lead. By analyzing the mobility coefficients (heavy metal growth in the Bt and Et layers), they can be ranked in the following decreasing sequence: on light soils: Cd > Cu > Zn > Pb and on medium soils: Cd > Zn > Pb > Cu.

Source of results: Own study and doctoral thesis entitled: "Influence of the selected physicochemical properties of the soil on the content of heavy metals in fodder plants" Warsaw University of Life Sciences, Department of Agriculture and Biology, Department of Agricultural Chemistry, 2006.

Author details

Dorota Pikuła
Department of Plant Nutrition and Fertilization, Institute of Soil Science and Plant Cultivation-State Research Institute in Puławy, Puławy, Poland

*Address all correspondence to: dpikula@iung.pulawy.pl

References

[1] World Health Organization (WHO). Science Communication Unit, University of the West of England, 2013. Science for Environment Policy. In-depth Report: Soil Contamination: Impacts on Human Health. Report produced for the European Commission DG Environment, September 2013. Available at: http://ec.europa.eu/science-environment-policy

[2] Alengebawy A, Abdelkhalek ST, Qureshi SR, Wang MQ. Heavy metals and pesticides toxicity in agricultural soil and plants: ecological risks and human health implications. Toxics. 2021;**9**(3): 42. DOI: 10.3390/toxics9030042

[3] Bolan NS, Mahimairaja A, Kunhikrishnan R. Naidu Sorption-bioavailability nexus of arsenic and cadmium in variable-charge soils. Journal of Hazardous Materials. 2013; **261**:725-732

[4] Wang X, Liu W, Li Z, Teng Y, Christie P, Luo Y. Effects of long-term fertilizer applications on peanut yield and quality and plant and soil heavy metal accumulation. Pedosphere. 2020; 30:555562. doi: 10.1016/S10020160(17) 60457-0. [CrossRef] [Google Scholar]

[5] Boussen S, Soubrand M, Bril H, Ouerfelli K, Abdeljaouad S. Transfer of lead, zinc and cadmium from mine tailings to wheat (Triticam aestivum) in carbonated Mediterranean (Northern Tunisia) soils. Geoderma. 2013;**192**:227-236

[6] Siuta J. Soil. Status and prognosis of land use and conservation in Poland. Journals Issues Progress of Agricultural. 1995;**418**(1):13-23

[7] Regulation of the Minister of Environment Protection on soil quality standards and ground quality standards (GJ No. 165/2002, pos. 1359) (in Polish)

[8] Ismael MA, Elyamine AM, Moussa MG, Cai M, Zhao X, Hu C. Cadmium in plants: uptake, toxicity, and its interactions with selenium fertilizers. Metallomics. 2019;**11**:255-277. DOI: 10.1039/C8MT00247A. [PubMed] [CrossRef] [Google Scholar]

[9] Miclean M, Cadar O, Levei EA, Roman R, Ozunu A, Levei L. Metal (Pb, Cu, Cd, and Zn) transfer along food chain and health risk assessment through raw milk consumption from free-range cows. International Journal of Environmental Research and Public Health. 2019;**16**(21):4064. Published online 2019 Oct 23. DOI: 10.3390/ijerph16214064

[10] Järup L. Hazards of heavy metal contamination. British Medical Bulletin. 2003;**68**(1):167-182 [PubMed] [Google Scholar]

[11] Rodriguez-Eugenio N, McLaughlin M, Pennock D. Soil Pollution: a Hidden Reality. Rome: FAO; 2018. p. 142

[12] Czarnowska K. Fertilization system and the content of Fe, Mn, Zn and Cu in rye plants at different developmental stages of rye. Soil Science Annual. 1993; **109**:67-76

[13] Kabata-Pendias A, Motowicka-Terelak T, Terelak H, Witek T. Assessment of the degree of contamination of soils and plants by heavy metals and sulfur—a framework of guidance for agriculture. Publisher Institute of Crop Science and Plant Cultivation, Puławy. 1993a;53:1–20. [Google Scholar]

[14] Kabata-Pendias A, Motowicka-Terelak T, Piotrowska M, Terelak H, Witek T. Assessment of the degree of

heavy metal and sulphur contamination of soils and plants. Framework guidelines for agriculture. IUNG, Puławy, P 1993b:53:7-10

[15] Terelak H, Motowicka-Terelak T, Stuczyński T, Pietruch Cz. Trace elements (Cd, Cu, Ni, Pb and Zn) in soils of agricultural land of Poland. Environmental Monitoring Library, Environmental Inspectorate, Warsaw, Poland, 2000, 69 pp. (in Polish)

[16] Klimkiewicz-Pawlas A. Environmental and health effects of soil contamination and methods for counteracting it. Studies and reports of Institute of Soil Science and Plant Cultivation, State Research Institute in Pulawy (Poland), 2021;66(20):91-116

[17] Perez A, Rodrigues-Eugenio N. Status of local soil contamination in Europe. JRC Technical Report, 2018

[18] Jaishankar M, Tseten T, Anbalagan N, Mathew BB, Beeregowda KN. Toxicity, mechanism and health effects of some heavy metals. Interdisciplinary Toxicology. 2014a;7(2):60-72. Published online 2014 Nov 15. DOI: 10.2478/intox-2014-0009

[19] Jaishankar M, Mathew BB, Shah MS, Gowda KRS. Biosorption of few heavy metal ions using agricultural wastes. Journal of Environment Pollution and Human Health. 2014b;**2**(1):1-6 [Google Scholar]

[20] Nagajyoti PC, Lee KD, Sreekanth TVM. Heavy metals, occurrence and toxicity for plants: a review. Environmental Chemistry Letters. 2010; **8**(3):199-216 [Google Scholar]

[21] Kabata-Pendias A. Soil-plant transfer of trace elements an environmental issue. Geoderma. 2004;**122**:143-149

[22] Najeeb U, Ahmad W, Zia MH, Malik Z, Zhou W. Enhancing the lead phytostabilization in wetland plant Juncus effusus L. through somaclonal manipulation and EDTA enrichment. Arabian Journal of Chemistry. 2014 [in press] [Google Scholar]

[23] Karaczun Z, Indeka L. Evironmental Protection. Warsaw: Publishing agency, Aries; 1996, 431 pp.

[24] Lamhamdi M, El Galiou O, Bakrim A, Nóvoa-Muñoz JC, Arias-Estévez M, Aarab A, et al. Effect of lead stress on mineral content and growth of wheat (*Triticum aestivum*) and spinach (*Spinacia oleracea*) seedling. Saudi Journal of Biological Sciences. 2013;**20** (1):29-36 [PMC free article] [PubMed] [Google Scholar]

[25] Sudhakar C, Syamalabai L, Veeranjaneyulu K. Lead tolerance of certain legume species grown on lead ore tailings. Agriculture, Ecosystems & Environment. 1992;**41**:253-261

[26] Brown G, Birkmann K. Heavy metals tolerance in Festuca opina L. Fromcontamited sites in the Eifel Mountains, Germany. Plant and Soil. 1992;**143**:239-247

[27] Steffens JC. The heavy metal – binding peptides of plants. Annual Review Plant Physiology Plant Molecular Biology. 1990;**41**:553-575

[28] Kabata-Pendias A. Trace Elements in Soils and Plants. Boca Raton, FL: CRC Press; 2010

[29] Hasselbach G. Ergebnisse zur Schwermetalltransfer Boden/Pflanze aufgrund von Gefässversuchen und chemischen Extraktionsverfahren mit Boden aus langjährigen Klärschlamm-Feldversuchen [dissertation]. Giessen: Justus Liebig Univeristät; 1992. p. 199

[30] Schaller A, Diez T. Pflanzenspezifische Aspekte der Schwermetallaufnahme und Vergleich mit den Richt - und Grenzwerten für Lebens - und Futerrmittel. Berichte aus der Ökologischen Forschung. 1991a;**6**: 94-114

[31] Kabata-Pendias A, Pendias H. Biogeochemistry of Trace Elements. PWN, Warsaw, 1993, 364 ss

[32] Sharma P, Dubey RS. Lead toxicity in plants. Brazilian Journal of Plant Physiology. 2005;**17**(1):35-52 [Google Scholar]

[33] Tchounwou PB, Yedjou CG, Patlolla AK, Sutton DJ. Heavy metal toxicity and the environment. Exp. Suppl. 2012;**101**: 133-164. DOI: 10.1007/978-3-7643-8340-4_6. PubMed Abstract | CrossRef Full Text | Google Scholar

[34] Gorlach E, Gambuś F. The effect of liming, adding peat and phosphorus fertilization on the uptake of heavy metals by plants. Polish Journal of Soil Science. 1991b;**2**:199-204

[35] Kubier A, Wilkin RT, Pichler T. Cadmium in soils and groundwater: a review. Applied Geochemistry. 2019; **108**:1-16. DOI:10.1016/j. apgeochem.2019.104388

[36] Krasowicz S, Górski T, Budzyńska K, Kopiński J. Agricultural characteristics of the area of Poland. In: Igras J, Pastuszak M, editors. The share of Polish agriculture in the emission of nitrogen and phosphorus compounds to the Baltic Sea, 2009, pp. 41-66. ISBN 978-83-7562-042-9

[37] Czarnowska K, Szymanowska-Sieńczewska W. Contents of soluble forms of Mn, Zn, Cu, Pb, and Cd in alluvial soils of the Vistula Valley and Żuławy. Soil Science Annual. 1999;**1**(2): 85-96

[38] Stępień W, Mercik S, Pikuła D. The influence of organic matter on the mobility of heavy metals in soil in a microplot experiment. Soil Science Annual. 2004a;**4**:149-156

[39] Stępień W, Mercik S, Pikuła D. The effect of organic matter on the mobility of heavy metals in soil in a microplot experiment. Soil Science Annual. 2004b; **4**:149-156

[40] Ciba J, Zolotajkin M, Cebula J. Chemical forms of cadmium in composts obtained from municipal waste contaminated with metallic cadmium dust and cadmium sulphate oxide. Arch. Ochr. Środr. 1998;**24**(2):131-140

[41] Chwastowska J, Skalmowski K, Sterlińska E, Wolska K, Skwara W. Heavy metals and forms of their occurrence in composts from urban waste according to the DANO technology in Warsaw. Archive Environmental Protection. 1993;**3-4**:251-259

[42] Wuana RA, Okieimen FE. Heavy metals in contaminated soils: a review of sources, chemistry, risks and best available strategies for remediation. ISRN Ecology 2011;**2011**2090–4614. Article ID 402647. DOI: 10.5402/2011/402647. CrossRef Full Text | Google Scholar

[43] Sarkar B. Heavy Metals in the Environment. New York-Basel: Marcel Dekker Inc.; 2002. p. 725

[44] Alloway BJ, Ayres DC. Chemical basis of environmental pollution. Transl. from English by S. Klosowicz, Warsaw: Publishing House PWN; 1999. 423 pp

[45] Piotrowska M, Teralak H. Cadmium in Polish soils. Journals Issues Progress of Agricultural. 1997a;**448b**:251-258

[46] Piotrowska M, Teralak H. Cadmium in soils of Poland. Journals Issues Progress of Agricultural. 1997b;**448b**: 251-258

[47] Zeien H, Brümmer GW. Ermittlung der Mobilitaet und Bindungsformen von Schermetallen in Böden mittels Sequentieller Extraktionen. Mitteilungen der Deutsche Bodenkundliche Gesellschaft. 1991;**66**(1):439-442

[48] Gorlach E, Gambuś F. Desorption and phytotoxicity of heavy metals depending on soil properties. Soil Science Annual. 1991a;**42**(3-4):207-214

[49] Ristovic S. Accumulation of Cd and Cr in alfalfa and lattuce plants gron on silos with different pH values. Review of Research Work AT the Faculty of Agriculture Belagrade. 1995a;**40**(2):81-90

[50] Ristovic S. Accumulation of Cd and Cr in lucerne and lattuce plants gron on silos with different pH values. Review of Research Work at the Faculty of Agriculture. 1995b;**40**(2):81-90

[51] Gębski M. Development of a method of fertilization on light soils contaminated with heavy metals [PhD thesis]. SGGW, Warsaw, 1997. 112 pp

[52] Fic M. Adsorptions und Desporptions – Verhalten von Cadmium, Chromium und Zink an – ausgewählten Böden und Sanden [PhD Thesis]. Christain Albrechts Univer., 1987, 99 ss

[53] Tangahu BV, Abdullah SRS, Basri H, Idris M, Anuar N, Mukhlisin M. A review on heavy metals (As, Pb, and Hg) up-take by plants through phytoremediation. International Journal of Chemical Engineering. 2011;**2011**: 939161

[54] Weber J. The influence of humus compounds on the accumulation and migration in soil of some heavy metals emitted by industry. Issue Notebooks of Agricultural Sciences. 1993;**411**:283-292

[55] Greinert H. The solubility and plant availability of heavy metals in silos contaminated with industrial dusts and sludges. In: Eijsackers HJP, Hamers T, editors. Integrated Soil and Sediment Research. A Basis for Proper Protection. Dordrecht: Kulwer; 1993. pp. 257-260

[56] Ratié R, Vaňková Z, Baragaño D, Liao R, Šípková A, Gallego JR, et al. Antagonistic Cd and Zn isotope behavior in the extracted soil fractions from industrial areas. Journal of Hazardous Materials. 2022;**439**:129519. DOI: 10.1016/j.jhazmat.2022.129519

[57] Ballabio C, Panagosa P, Lugatoa E, Huang JH, Orgiazzia AH, Jonesa A, et al. Copper distribution in European topsoils: an assessment based on LUCAS soil survey. Science of the Total Environment. 2018;**636**:282-298

[58] Kabata-Pendias A, Pendias H. Trace Elements in Soils and Plants. Boca Raton: CRC; 2001 [Google Scholar]

[59] Gąszczyk R, Paszko T. A study on the stability of soil sorption complex connections with Cu, Ni, Zn by a dynamic method. Journals Issues Progress of Agricultural. 1995;**418**(part II):495-500

[60] Tukiendorf A. The Role of Metalloprotein Complexes in the Tolerance of Higher Plants to Toxic Concentrations of Heavy Metals. Lublin, Dissertation: UMCS; 1990, 70 pp

[61] Hamilton EI. The periodic table of the elements: geochemical and biochemical associations. In: Coughtrey PJ, Martin MH, Unsworth MH, editors. Pollutant Transport and Fate in

Ecosystems. Oxford: Blackwell Sci. Publ; 1987. pp. 5-33

[62] Król A, Mizerna K, Bożym M. An assessment of pH-dependent release and mobility of heavy metals from metallurgical slag. Journal of Hazardous Materials. 2020;**384**:121502

[63] Horckmans LR, Deckers JS. Geochemical and mineralogical study of a site severely polluted with heavy metals (Maatheide, Lommel, Belgium). Environmental Geology. 2006;**50**:725-742. DOI: 10.1007/s00254-006-0245-x

[64] Dudka S. Evaluation of the total contents of major and trace elements in the surface layer of the layer of Polish soils. Pulawy: IUNG, R(293); 1992

[65] Herms U, Brümmer GW. Einflussgrössen der Schwermetallöslichkeit und bindung in Böden. Boden und Pflanzenernährung. 1984;**147**:400-424

[66] Hiller DA, Brümmer GW. Mikrosondenuntersuchnungen zur Bestimmung der Bindunsformen von Schwermetallen in Böden. Berichte aus der Ökologischen Forschung. 1991;**6**:36-58

[67] Gworek B, Kozera-Sucharda B. Zeolites- genesis, structure and basic physical properties. Environmental Monitoring and Assessment. 1999;**17**: 157-168

[68] Czuba R, Ciećko Z, Cwojdziński W, Gorlach E, Grzebisz W, Grześkowiak A, Kamiński E, Klupczyński Z, Kostuch R, Krzywy E, Masztalerz P, Mazur T. Mineral fertilization of cultivated plants. Chemical Works Publishers "Police". 1996, pp. 315-368

[69] Fischer WR. Die Komplexierung von Cu, Pb und Cd durch heisswasserlösliche Huminstoffe. II Die Stabilität der gebildeten Komplexe. Boden und Pflanzenernährung. 1986;**149**(1):120-129

[70] FAO, World reference base for soil resources, by ISSS–ISRIC–FAO. World Soil Resources Report No. 84; (1998). Rome, 88 pp

[71] Gembarzewski H, Korzeniowska J. The use of 1 M HCl solution as a common extractant to assess the abundance of soil in bioavailable forms of micronutrients. Prace Kom. Nauk. PTG no 99, Warsaw, 1987. pp. 1-9

[72] Starck Z. physiological basis of plant productivity. In: Kopcewicz J, Lewak S, editors. Physiology of Plants. Warsaw: PWN Publishing House; 2002. pp. 679-706

[73] Piesak Z. Analysis of the possibilities of agricultural use of areas degraded as a result of the impact of industry in the province. Katowice. Th. I. Possibilities of using the areas located in the central part of the province. Katowice for the production of feed. Environmental protection bulletin, Katowice, 1986b, (Typescript, 20 pp.)

[74] Piesak Z. Analysis of the possibilities of agricultural use of areas degraded as a result of the impact of industry in the province. Katowice. Th. I. Possibilities of using the areas located in the central part of the province. Katowice for the production of feed. Bull. IOŚ – SOMETHING, Katowice, 1986b, (Typescript, 20 pp.)

[75] Curyło T, Jasiewicz C. Comparison of the effects of mineral and mineral-organic fertilizers on the uptake and toxicity of cadmium and nickel to plants. Journals Issues Progress of Agricultural. 1997;**448a**:45-52

[76] Chłopecka A. Influence of Different Compounds of Cd, Cu, Pb and Zn on the Forms of These Metals in Soil and on Their Content in [Cereal] Plants. Vol. R (315). Puławy: IUNG; 1994. p. 65

[77] Wallace A, Romney EM. Root of highter plants as a barier to translocation fo some metals to shoots plants. In: Proceedings of the 15 th Annual Handford Life Sciences Symposium, 1975. pp. 370-379

[78] Gorlach E. Plant as a mitigating factor for the negative effects of heavy metal pollution of soils. Regional Bulletin of the Academy of Agriculture, Kracow. 1994;**305**:163-172

[79] Horubala A. Food contamination. Food Industry. 1991;**8**:189-190

[80] Brümmer GW, Gerth J, Herms U. Heavy metal species, mobility and availability in soils. Zeitschrift für Pflanzenernährung und Bodenkunde. 1986;**149**:382-389

[81] Strączyński SJ. Copper content in selected plant species grown on copper-contaminated soils. Journals Issues Progress of Agricultural. 2000a;**472**:619-662 Ch. II

[82] Strączyński SJ. Copper content of selected plant species grown on copper-contaminated soils. 2000b;**II**(472):619-662

[83] Kucharski R, Marchwińska E, Piesak Z, Nikodemska E, Witała B. Contamination of feed plants with lead and cadmium in selected regions of the voivodeship Katowicki. Mediterian Wet. 1984;**3**:162-165

[84] Chaudhry FM, Sharit M, Latif A, Querreshi RH. Zinc-copper antagonism in the nutrition of rice. Plant and Soil. 1973;**38**:573-580

[85] Traynor MF, Knezek BD. Effekt of nickel and cadmium contaminated soils on nutrient composition of corn plants. Trace Substances, Environment and Health. 1973;7:83-87

[86] Lagerwerff JV. Current research in heavy metals in soil, sediment and water. In: Proc. 2nd Ann. Conf. Trace Contamin., NSF. Pacific Grove, CA, 1974. pp. 16-47

[87] Sapek A, Sapek B. Factors determining the content of cadmium and lead in grassland forage. Handbook of the Institute of Land Improvement and Grassland; 1989a, pp. 149-150.

[88] Sapek A, Sapek B. Determinants of cadmium and lead content in grassland forage. Handbook of the Institute of Land Improvement and Grassland; 1989b, pp. 149-150

[89] Gębski M, Mercik S, Stępień W. Application of chemical extraction methods to predict the degree of heavy metal contamination of soils and plants. Final report of the research project to the State Committee for Scientific Research no. 6 PO4G045 15. Department of Soil Science, Warsaw University of Life Sciences, Warsaw, 2002. 40 pp

[90] Warda M. Influence of soil properties on cadmium and nickel accumulation in grasses and dicotyledonous plants selected from a pasture sward. Journals Issues Progress of Agricultural. 1997;**448**(a): 347-351

[91] Greger M, Johanson M. Cadmium effects on leaf transpiration of sugar beet (Beta vulgaris). Pysiology Plant. 1992;**86**: 465-473

[92] Kabata-Pendias A, Piotrowska M, Motowicka-Terelak T, Maliszewska-Kordybach B, Filipiak K, Krakowiak A,

et al. Grounds for assessment of chemical contamination of soils. Heavy metals, sulfur, PAHs. Warsaw: Library of Environmental Monitoring; 1995 [Google Scholar]

[93] Piotrowska M, Kabata-Pendias A. Impactof soils amended with Zn and Pb smelter dust on Cd concentrations in potatoes. Journal of Geochemical Exploration. 1997;**58**:319-332

[94] Kabata Pendias A, Pendias H. Biogeochemistry of Trace Elements. Warsaw: PWN; 1999. p. 398

[95] Łyszcz S, Ruszkowska M. Differential response of several plant species to excess zinc in soil. Soil Science Annual. 1991;**42**(314):215-221

[96] Cherfi A, Abdoun S, Gaci O. Food survey: levels and potential health risks of chromium, lead, zinc and copper content in fruits and vegetables consumed in Algeria Food Chem. Toxicology. 2014;**70**:48-53

[97] Falkowski M, Kukułka I, Kozłowski S. Chemical properties of meadow plants. Wyd. AR Poznań, 1990. 45 ss

[98] Gorlach E, Gambuś F. A comparison of sensitivity to the toxic action of heavy metals in various plant species. Polish Journal of Soil Science. 1992;**2**:207-213

[99] Regulation of the Minister of Agriculture and Rural Development of 23 January 2007 on the permissible content of undesirable substances in feed. Dz. U. 2007, No. 20, item 119 (in Polish)

[100] Niemyska-Łukaszuk J. The influence of granulometric composition and soil reaction on the content of bioavailable forms of heavy metals. Journals Issues Progress of Agricultural. 1995;**418**:459-464

[101] Dombrowski N. The influence of physicochemical properties of soils and the degree of their contamination with heavy metals on the uptake of these elements by potatoes [Master's thesis]. Warsaw University of Life Sciences, Warsaw, 2001, pp. 7-41

[102] Mercik S, Stępień W, Gębki M. Cadmium content in plants and in the soil profile as a result of two or three applications of lime from a lead smelter. Journals Issues Progress of Agricultural. 1997;**448a**:253-259.

[103] Cyganov A, Vihldflush I, Kal M. Study on detoxification methods of sward–podzolic light loamy silos polluted with heavy metals. Journals Issues Progress of Agricultural, Issu. 2001;**476**:91-98

[104] Fijałkowski K, Kacprzak M, Grobelak A, Placek A. The influence of selected soil parameters on the mobility of heavy metals in soil. Environment Protection Engineering. 2012; **15**(1):81-92

[105] Bowen HJM. Environmental Chemistry of the Elements. London: Academic Press; 1979. p. 333

[106] Vaněk A, Borůvka L, Drábek O, Mihaljevič M, Komárek M. Mobility of lead, zinc and cadmium in alluvial soils heavily polluted by smelting industry. Plant, Soil and Environment. 2005;**51**(7): 316-332

[107] Gruszecka A, Wdowin M. Characteristics and distribution of analyzed metals in soil profiles in the vicinity of a postflotation waste site in the Bukowno region, Poland. Environmental Monitoring and Assessment. 2013;**185**(10):8157-8168. Published online 2013 Mar 22. DOI: 10.1007/s10661-013-3164-9

[108] Gruszecka A, Helios-Rybicka E. Distribution of Zn and Pb in the vicinity of non-ferrous industrial waste sites at the example of Bukowno (Poland) and Mansfeld (Germany) Polish. Journal of Environmental Studies. 2006;**15**(5c): 164-170 [Google Scholar]

[109] Kaniuczak J, Hajduk E. Cadmium and lead in some soils of South-Eastern Poland. Zesz. Probl. Post. Nauk Rol. 1995;**418**(1):241-246

[110] Regulation of the Minister of Agriculture and Rural Development of 28 June 2004, on the permissible content of undesirable substances in fodder (Journal of Laws No. 162, item 1704)

Chapter 7

Investigating Bacterial Communities Resistant to Heavy Metal and PAHs Pollutants in the Persian Gulf and Their Ability to Remove These Compounds (a Review)

Razieh Lamoochi and Kobra Jalali

Abstract

The Persian Gulf is one of the most diverse water environments in the world. There are various types of marine creatures including corals, sponges and fish in this marine environment. These pollutant sources in the Persian Gulf include oil spills from oil tanks, shipping accidents, marine transportation and oil extraction processes. Since nearly 60% of the world's oil is transported to the Persian Gulf, oil pollution is inevitable. This has made the marine environment the most polluted sea in the world. Bioremediation can be defined as the removal of pollutants such as heavy metals and PAHs from solution by biological agents such as bacteria, fungi, microalgae and yeasts. Various species of bacteria have been isolated and reported by several researchers in the Persian Gulf, perhaps due to its high resistance to a wide range of petroleum hydrocarbons and heavy metals. Therefore, in this chapter, we decided to review the studies conducted in the field of isolating and identifying native bacteria and evaluating their ability to remove heavy metals and polycyclic aromatic hydrocarbons in the Persian Gulf.

Keywords: Persian gulf, PAHs, heavy metals, bioremediation, bacteria

1. Introduction

The rapid growth of industrial activities in recent years and non-obedience with environmental laws accompanied by discharging of various pollutants from petrochemical industries cause increase of contaminants into the environment [1, 2]. Wastewater comprises many organic and inorganic pollutants, and settling them into received waters causes to serious environmental problems [3]. The entrance of huge amounts of contaminants such as oil compounds and heavy metals via the wastewaters of industries to the coastal and aquatic ecosystems has created serious problems [4, 5].

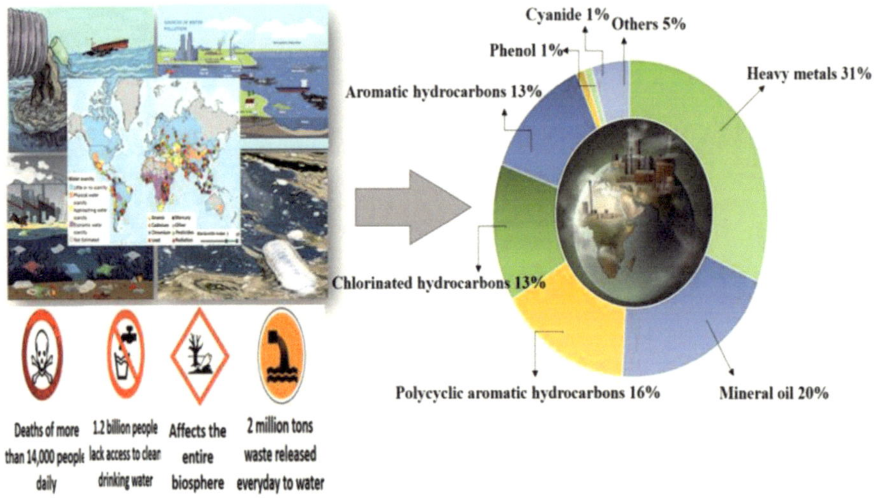

Figure 1.
Summary of global water pollution worldwide and geographical distribution of water pollution based on the type of main pollutant [5].

The presence of pollutants in water bodies above their acceptable levels set by the World Health Organization (WHO) and Environmental Agencies may have enormous effects [6]. Most of wastes discharged to the environment contain toxic contaminants such as aromatic compounds and heavy metals which can cause major damage to marine ecosystems and living organisms such as neuronic toxicity, carcinogenicity and reproduction abilities [7, 8]. Therefore, it is necessary to reduce the levels of pollutants below the standard range (**Figure 1**) [9].

Many methods have been offered for treatment of contaminated water environments, but they may be non-effective or expensive. However, bioremediation is an effective technology with a range of advantages rather than traditional methods [10, 11]. Bioremediation of waste materials which contain hydrocarbons and heavy metals is based on the microorganism's potential such as bacteria and fungi to absorb them from polluted wastewater or change them to non-toxic products [11–13].

The microorganisms could be native to a polluted area or isolated from another place and moved to the contaminated position. PAHs and heavy metals in the environment can influence microbial communities and form new microbial communities with the purpose of adapt to the hazardous environment so the strains screened from these contaminated areas usually show high capacity to tackle with the pollution of PAHs and heavy metals [11, 14].

2. Type of pollutants

2.1 Polycyclic aromatic hydrocarbons (PAHs)

Polycyclic aromatic hydrocarbons (PAHs) are a group of common environmental contaminants that are especially widely detected in the aquatic environment categorized as general environmentally harmful pollutants [15]. PAHs typically enter the environment through the natural sources and anthropogenic sources. Examples of

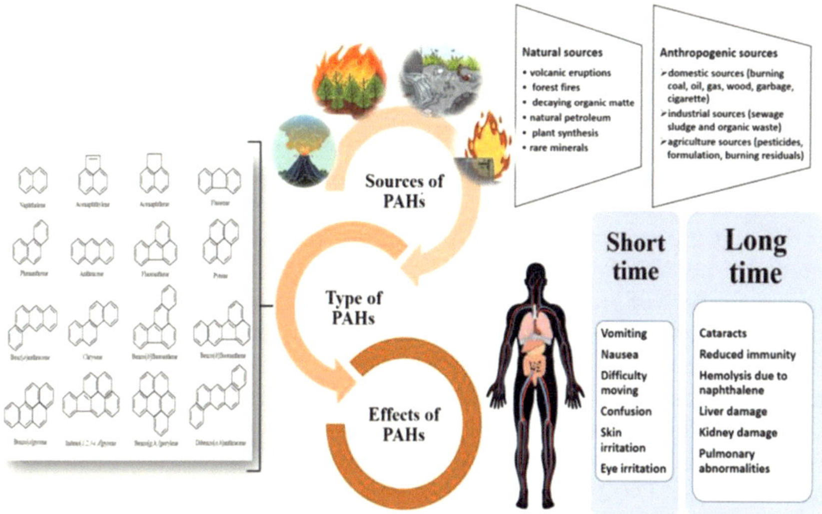

Figure 2.
Sources of PAHs, types of PAHs and their effects on humans.

natural sources of PAHs formation include volcanic eruptions, forest fires, decaying organic matter, natural petroleum, plant synthesis, rare minerals, anthropogenic sources of PAHs such as domestic sources (burning coal, oil, gas, wood, garbage and cigarette), industrial sources (sewage sludge and organic waste), and agriculture sources (pesticides, formulation, and burning residuals) [3, 16, 17].

PAH compounds contain only carbon and hydrogen atoms. Chemically the PAHs are comprised of two or more benzene rings bonded in linear, cluster or angular arrangements [17]. PAHs form up to six fused aromatic rings are often referred to as small PAHs, and those containing more than six aromatic rings are called large PAHs [3, 17]. Such molecular arrangements are illustrated in **Figure 2**.

PAHs are hydrophobic compounds that degrade hardly in water and sediment. Due to their hydrophobicity, they are transferred into cells and induce gene expression of the cytochrome P450 (CYP) enzyme group [18, 19]. They have been shown to have carcinogenic and mutagenic effects and are strong immune suppressants. In addition, toxicological studies have revealed other types of PAH toxicities such as developmental toxicity, genotoxicity, immunotoxicity, oxidative stress and endocrine disruption. Because of their ubiquity in the natural environment and various harmful effects on organisms, PAHs are among the most concerning organic pollutants, and The Environmental Protection Agency has introduced these compounds as hazardous contaminants [20, 21].

2.2 Heavy metals

Heavy metals (HMs) are considered one of the most important inorganic pollutants in marine environments which the extensive use and distribution of them has increased rapidly all over the world [22]. They are usually released through human activities such as industrial operations, agriculture activities, mining, electroplating, electronic waste/E-waste and geological sources include volcanic eruptions, weathering of metal-containing rocks, sea-salt sprays, forest fires and natural weathering processes [22, 23].

Among heavy metals magnesium (Mg), nickel (Ni), chromium (Cr3+), copper (Cu), calcium (Ca), manganese (Mn), and sodium (Na) along with zinc (Zn) are vital elements needed in slight quantity for metabolic and redox functions [24]. They act as co-enzymes for catalytic enzymes in metabolic cell reactions, but increasing their concentration can lead to toxic condition and cause numerous carcinogenic effects [24, 25]. On the other heavy metals for instance aluminum (Al), lead (Pb), cadmium (Cd), gold (Au), mercury (Hg) and silver (Ag) do not have any biological role and are toxic to living organisms [24]. They are able to prevent the regular physiological and developmental functions of living organisms in minor amounts and responsible for a wide range of physiological diseases [24, 26, 27].

Heavy metals are of greater concern among the uncountable organic and inorganic compounds released into the aqueous environments due to their persistence, potential of bioaccumulation and toxicity in various species [24, 28].

They show pronounced toxicity effects in terms of bioaccumulation and biomagnification in tissues and food webs, respectively. Increases in heavy metal concentration, to levels higher than standard amounts identified, may lead to sub-lethal or even lethal impacts on aquatics and humans [24, 26, 28]. When heavy metals are taken up in higher concentrations, they are be responsible for posing serious health risks and can affect liver, brain, bone and lungs, although each metal has its own particular signs and symptoms [25, 26].

The key mechanism of heavy metal toxicity comprises the production of free radicals to cause oxidative stress, harm of biological molecules such as enzymes, proteins, lipids and nucleic acids, damage of DNA which is important to carcinogenesis besides neurotoxicity [24, 29]. Some heavy metal toxicity can be acute, while others can be chronic after long-term exposure which may lead to the damage of numerous tissues in the body for example the brain, lungs, liver and kidney causing diseases in the body (**Figure 3**) [30].

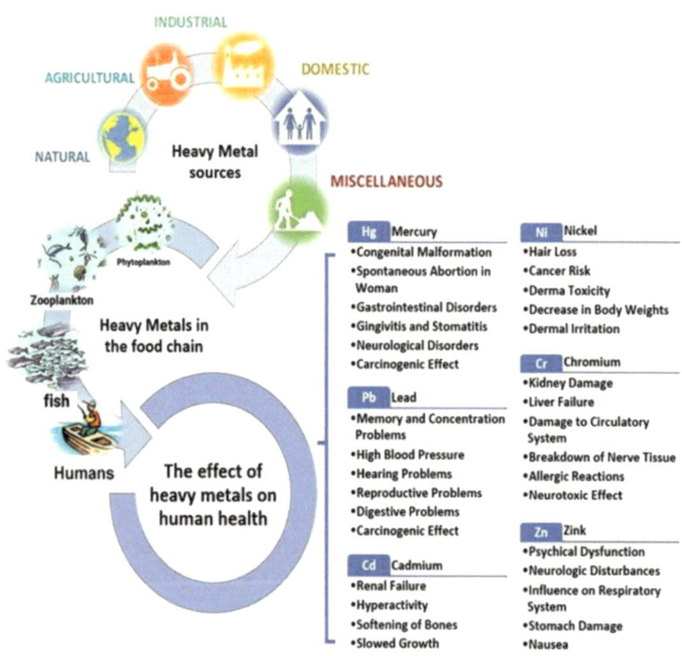

Figure 3.
Heavy metals: Sources, mechanism of action and its effects on humans.

3. Biological treatment

For the reason that PAHs and HMs cause serious problems in environment even at little concentrations, elimination of these compounds has become an urgent obligation in recent decades, and it is essential to develop techniques for their removal [31, 32]. Many methods are being used for treatment of contaminated water environments. Pollutants can be removed using various treatment technologies, including chemical precipitation, oxidation or reduction, filtration, ion-exchange, reverse osmosis, membrane technology, evaporation and electrochemical treatment [33, 34]. However, these methods are always not effective. Moreover, they are expensive and relatively ineffective, and they can generate extra problems. Consequently, they are seen as unfavorable and uneconomical [35, 36].

In recent years natural treatment systems are regarded as the best available technology for treating various types of wastewaters. Nowadays experts have turned to a new method for removing pollutant from the environment using microorganisms [37, 38]. Removal of heavy metals and PAHs from aqueous solution by using microorganisms is called bioremediation which are considered to be the more appropriate than the physical and chemical methods because of their high removal and recovery efficiency, cost-effective and reducing the production of secondary toxic compounds [35].

Bioremediation is defined as the process whereby organic and inorganic wastes biologically become under controlled conditions to an innocuous state, or to levels below concentration limits established. Since the early 1900s, biological biomass used to diminish nutrients, pollutants and organic matters in wastewater [33, 35].

The mechanisms, such as biosorption, bioaccumulation and biodegradation, are involved in bioremediation. Both bioaccumulation and biosorption are the processes using which microbes or biomass gets bound to the HMs and pollutants from the surrounds and concentrates them [35, 36, 38].

Biosorption is a process in which microorganisms use their cellular structure to capture HM ions, which they then absorb onto the cell wall's binding sites. It is a passive uptake process and does not depend on the metabolic cycle [30]. Biosorption is a reversible and fast passive adsorption process, and there may be both living and dead biomass for biosorption [9, 38].

However, unlike biosorption, bioaccumulation by microbes is an active metabolic process in which heavy metals accumulate and are transported into intracellular living bacterial cells using proteins. Endocytosis, ion channels, carrier-mediated transport, complex permeation and lipid permeation are all involved in HM bioaccumulation in the bacterial membrane [39]. This process depends on microorganism cell metabolism. In contrast to biosorption where dead biomass cells are still able to eliminate HMs, bioaccumulation occurs only by living biomass cells [40].

In biodegradation, microorganisms degrade toxic organic pollutants into safe and non-toxic compounds. In fact, biodegradation basically occurs due the actions of multiple organisms which enzymatically deal with the pollutants and turn them into innocuous products [41, 42].

Using biological techniques for eliminating of heavy metals and aromatic compounds from marine ecosystems is considered reasonable. There are number of advantages associated with the use of microorganisms to removal of pollutants (**Figure 4**) [43, 44].

Bacteria, microalgae, fungi and yeast are the most important microorganisms that are commonly used for bioremediation [45–47]. They are extensively dispersed on the environment because of their significant metabolic capacity and they can simply

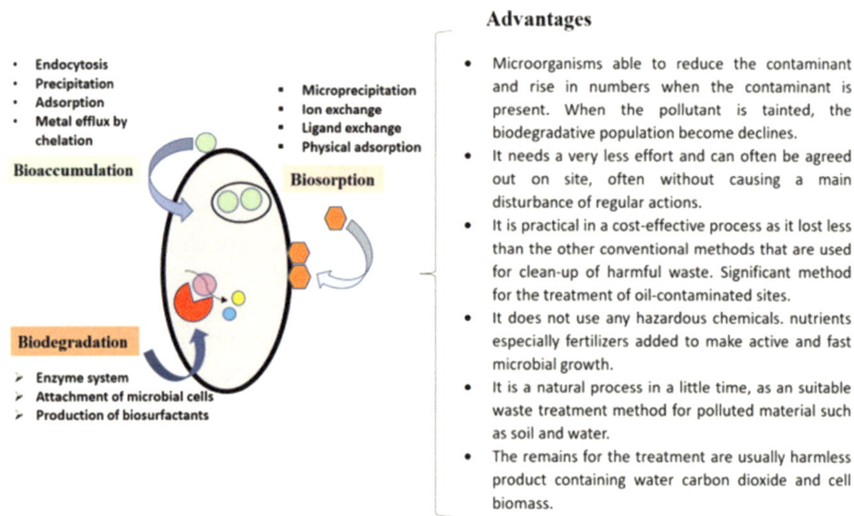

Figure 4.
Types of bioremediation mechanisms and their benefits.

grow in a wide range of environmental conditions [43]. However, compared to other microorganisms, bacteria signified greater removal ability due to their small size, high rate of growth and reproduction and active biosorption sites such as peptidoglycan. On the other hand, bacteria are considered the most exceptional bio-sorbents among all other organisms because of their high surface-to-volume ratios and active sites of chemosorption in their cell wall, such as teichoic acid [48, 49].

4. Persian gulf

Around 70% of the earth is enclosed by sea water. Oceans provide food, recreation and transportation that sustain a significant portion of the world's economy [50]. The Persian Gulf is one of the most diverse water environments in the world [51]. There are various types of marine creatures including corals, sponges and fish in this marine environment [52, 53]. Mangrove forests are found in this aquatic ecosystem. In mangrove forests, some of the complex ecological relations between organisms living in intertidal zones [54]. Other marine creatures in the Persian Gulf are also very diverse. In recent years, because of human activities, various contaminants have entered the oceans and marine environments, leading to the change of life in these main aquatic ecosystems [55, 56]. The Persian Gulf is one of the main ways of transferring energy and goods. Since about 60% of the world's oil is transported to the Persian Gulf, oil pollution is unavoidable [57, 58].

In the countries of the Persian Gulf, air, soil and water pollution by mineral and organic pollutants are an important subject. Also, during the last 30 years, there have been three wars in the Persian Gulf: The Islamic Republic of Iran and Iraq War (1980–1988), the first and second Persian Gulf Wars (1991 and 2003, respectively) [59]. Thus, in 1991, a large oil spill occurred in the Persian Gulf, and as a consequence, 6 to 8 million barrels of Kuwaiti crude oil were discharged into the Persian Gulf [60]. The oil spill due to the first Persian Gulf War is considered the worst oil spill in history. Therefore, a large amount of research was focused on the same spill, which

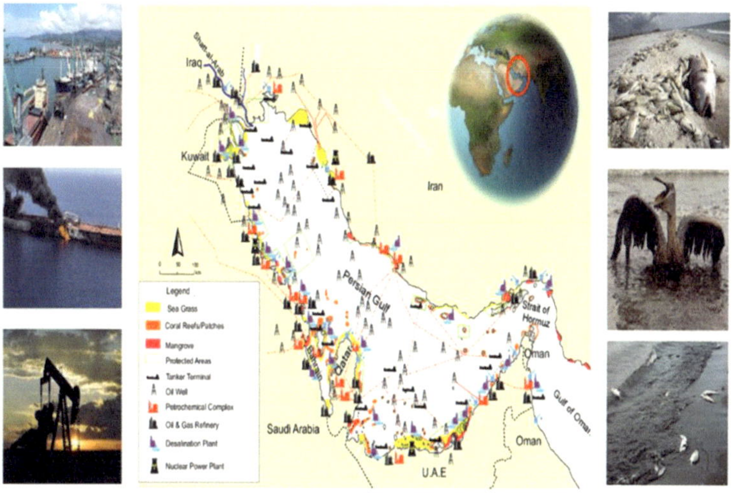

Figure 5.
Main environmentally sensitive parts of the Persian Gulf, key oil and gas infrastructures, chief ports, desalination plants and nuclear energy facilities [57].

confirmed the range of influence of the oil spill of 400 km from the point where the spill occurred to the coastlines of Saudi Arabia [3, 59, 60]. Moreover, numerous spills produced by routine oil operations and tankers. So, The Persian Gulf has experienced numerous and enormous pollution [61]. On the other hand, reducing the distribution and dispersal of pollutants because of the shallow depth and limited contact of the Persian Gulf with other aquatic ecosystems causes contaminants to remain in the Persian Gulf for a long time [62, 63].

These factors have turned the Persian Gulf into one of the most polluted seas in the world. Oil pollution in the Persian Gulf has not been managed, many corals, sponges. Mangrove forests are soon destroyed, and biodiversity is significantly reduced. Therefore, the protection of this water environment is more vital in the world (**Figure 5**) [63–65].

Many studies have been conducted to remove heavy metals and PAHs from aquatic environment using biological adsorbents, which are considered the most appropriate methods compared to other methods [66, 67]. Overall, various species of bacteria have been isolated and reported by numerous researchers in the Persian Gulf, maybe because of its high resistance to an extensive range of oil hydrocarbons and heavy metals [3, 6, 68]. According to mentioned abilities of bacteria, many writers have suggested the use of this method in removing oil pollution and heavy metals in aquatic environments such as the Persian Gulf [69–74].

In this chapter, we decided to have a review of the studies done on isolation and identification of indigenous bacteria and evaluation of its ability in removal of heavy metals and polycyclic aromatic hydrocarbons in Persian Gulf.

5. A review of studies on bioremediation in the Persian Gulf

Marine ecosystems around the world, such as the Persian Gulf, are suitable habitats for all kinds of native bacteria [75]. These bacteria belong to different genera that are mostly found in the aerobic zone of marine sediments [71, 76]. In the last two decades, many researches have been conducted on non-symbiotic bacteria and their performance in cleaning the Persian Gulf [3, 6, 77–81].

Numerous hydrocarbon-utilizing bacteria have been isolated from the Persian Gulf which Proteobacteria, for example, Gammaproteobacterial genera (e.g., Acinetobacter, Macrobacteria and Alcanivorax), Alphaproteobacterial genera (e.g., Tissrella and Zavarzina) and also Actinobacteria are the dominant cultivatable marine bacteria [82–84].

Studies have shown that marine bacteria often belong to the genera of Marinobacter, Microbacterium, Rhodococcus, Kocuria, Alcanivorax, Microbacterium, Pseudomonas, Pseudoalteromonas, Dietzia and approximately all of them are found in the aerobic zone of marine sediments. Besides the bacteria referred above, Heretofore, various hydrocarbon-degrading bacteria have been caught from Kish Island in Iran. The most commonly reported genera of hydrocarbon-degraders include Pseudomonas, Acinetobacter, Nocardia, Vibrio and Achromobacter.78 s [78, 85–88].

Likewise, Hassanshahian et al. separated bacteria resistant to various petroleum compounds from the Persian Gulf. In their research, they succeeded in isolating 25 oil-resistant bacteria from the sediments of the Persian Gulf and the Caspian Sea. Molecular identification showed that these bacteria belong to these genera: Acinetobacter, Cobetia, Gordonia, Rhodococcus, Pseudomonas, Halomonas, Microbacterium, Marinobacter and Alcanivorax. Their results showed that the studied bacteria can be used to remove pollutants from the environment and reduce the level of oil pollution in the Persian Gulf and the Caspian Sea [89].

In the other study, *Rhodococcus ruber* KE1 was isolated from Persian Gulf water and sediment (khark Island, south part of Iran), and the ability of this bacterium to break down petroleum compounds was evaluated. Different experimental parameters such as pH, temperature, crude oil concentration and agitation speed were optimized to achieve the maximum degradation efficiency. The best results were measured at pH 8.5, temperature 40°C, stirring speed 250 rpm and crude oil concentration 1% (v/v). Associated data showed that *R. ruber* KE1 under optimal conditions is able to remove about 90% of the total hydrocarbons after one week. The measurement of residual nitrogen and sulfur showed a decrease of 48 and 44%, respectively. The ability of this particular strain as a surfactant producer was measured under optimal conditions. A decrease in surface tension was recorded from 60 (control) mN/m to 26 mN/m [90].

In the study of Kafilzadeh et al., sampling of mangrove sediments was done. After bacterial counting and enrichment in MSM medium, the isolated bacteria were purified on blood agar. Bacteria were identified by different biochemical tests and PCR. *Brevibacterium iodinum*, *Acinetobacter calcoaceticus*, *Corynebacterium variabile*, Enterobacter sp. and Acinetobacter lowffi were isolated from mangrove sediments. Consequences displayed 86.2, 81.4, 75.2, 57.3 and %55.8 of substrate was degraded by these bacteria respectively. Based on results of growth studies, *B. iodinum*, *A. calcoaceticus* and *C. variabile* showed the maximum growth after 72 hours, and Enterobacter sp. and Acinetobacter lowffi had the maximum growth after 84 hours of incubation [91].

However, based on other studies conducted in the northern coasts of the Persian Gulf, more different species of bacteria resistant to heavy metals and petroleum hydrocarbons were found [6, 77, 92–95]. One of the most important bacterial species isolated from the sediments of the Persian Gulf is the species belonging to the genus Pseudomonas [77, 92–95].

Among ten bacterial species isolated from marine sediment, by Safahieh et al., one strain represented high potential to grow in medium supplemented with copper and phenanthrene. Isolated bacterium was identified as Pseudomonas sp. by biochemical tests. Over 70% of copper absorbed on Pseudomonas sp. within 150 min. In addition, about 96.52% of primary concentration of phenanthrene was degraded during 120 h. Based on the capacity of isolate in degradation of phenanthrene and biosorption of copper, this species is introduced as a suitable microorganism for bioremediation of polluted environments [77].

Also, Lamoochi et al., in another study, *Pseudomonas aeroginosa* was isolated and identified as a resistant bacterium to zinc from Musa Bay sediments (located in the North of the Persian Gulf, Iran). Investigating the growth of *P. aeroginosa* bacteria in an aqueous solution containing 20, 40, 80, 160 and 320 mg/L of zinc showed that increasing the concentration of zinc produced a decrease in bacterial growth. The maximum bacterial growth of 0.93 was achieved at a concentration of 20 mg/L zinc. By increasing the zinc concentration to 320 mg/L, bacterial growth reduced to 0.37, which showed an important difference with the maximum bacterial growth at 20 mg/L zinc concentration. *P. aeroginosa* can remove 86% of Zn at a concentration of 25 mg/L Zn [92].

In order to compare the growth and ability of biosurfactant production in bacteria present in oil-contaminated sediments of the Persian Gulf (Khor Musa), samples were taken from the contaminated sediments of the region. Two species of bacteria Psedomonas stutzeri and *Alcaligenes denitrificans* were isolated and purified from oil contaminated sediments and identified as oil-resistant bacteria through morphological and biochemical tests. Bacterial growth was measured by a spectrophotometer at optical density at a wavelength of 600. Evaluation of biosurfactant was done with blood agar, drop collapse and oil spread methods. Psedomonas stutzeri showed that it is able to produce biosurfactant. According to results, Psedomonas stutzeri has a good growth in the media comprising Antheracene as single source of carbon and energy. Furthermore, the data showed that the growth of biosurfactant-producing bacteria (Psedomonas stutzeri) was better than *Alcaligenes denitrificans* due to the increased solubility of anthracene and its increased bioavailability in the environment [93].

Uptake of heavy metal ions by *Pseudomonas aeruginosa* strain MCCB 102 isolated from the Persian Gulf was investigated in two single and multi-mix forms. The highest adsorption was observed for Cu, Zn, Cd and Pb, respectively. Removal of metals was maximized when the metal ions were in a single form. Scanning electron microscope (SEM) analysis showed *P. aeruginosa* strain MCCB 102 accumulated heavy metals in the cell wall and along the external cell surfaces [96].

A study by Shahaliyan et al. (2016) aimed to isolate and identify the species *Ochrobactrum anthropi* and *Pseudomonas putida* as indicators for measuring lead absorption and degradation of anthracene in laboratory conditions. The consequences presented that the mentioned species could grow well in 50 mg/l of lead and 30 mg/l of anthracene. The maximum growth at concentrations of 50 mg/l of lead and 30 mg/l of anthracene was measured 1.39 and 0.530, respectively. The results of the study of the capability of lead uptake by *O. anthropi* at a concentration of 50 mg/l and anthracene elimination by P. putida at a concentration of 30 mg/l indicated that the mentioned bacteria started absorbing lead and anthracene at the time of inoculation and rapidly reduced their concentrations in the solution. Elimination of lead and anthracene by *O. anthropi* and *P. putida* from the metal and hydrocarbon solutions continued actively till the last moments of measurement and at the end of the experiment, 84% of lead and 77.986% of anthracene were removed [95].

Pseudomonas Sp. is a saprotrophic bacterium with adaptability to various environments [77, 96, 97]. This bacterial species has diverse aerobic metabolisms for bioremediation of aromatic compounds and biosorption of heavy metals. It appeared that it has the ability to use benzene, toluene and naphthalene as the sole source of carbon and energy. The ability of Pseudomonas species in degradation of hydrocarbon and heavy metal materials has been reported by many researchers [98, 99].

Also, various species of Bacillus have been isolated and reported by several investigators possibly due to its high resistance to wide range of oil hydrocarbons and heavy metals According to previous studies, the genus Bacillus has a significant ability to remove heavy metals and PAHs [6, 100–103]. This genus comprises gram-positive, aerobic and spore-producing bacteria with the ability to absorb heavy metals due to special binding sites in their cell walls (such as teichoic and teichronic acids) [6, 104–106].

In another study, native bacteria resistant to lead were isolated from the surface sediments of Imam Khomeini Port (Persian Gulf). The isolated bacterium *Bacillus*

Bactrial name	Isolation area	pollutant	Reference
Rhodococcus ruber KE1	Persian Gulf	Crude Oil	[90]
Pseudomonas putida strain 1389	Persian Gulf	Naphthalene	[110]
Amphibacillus xylanus *Bacillus pasteurii*	Imam Khomeini Port Persian Gulf	Anthracene	[107]
Pseudomonas sp.	Khor Musa, Persian Gulf	Phenanthrene	[77]
Bacillus subtilis	Persian Gulf	Anthracene	[6]
Halomonas aquamorina *Gordonia bronchialis* *Flavobacterium* *Gaetbulibacter* *Tistrella*	Persian Gulf	hydrocarbon	[79, 82, 109]
Nocardia *Achromobacter* *Sphingomonas* *Halomonas sp. MS1*	Kish Island, Iran	Crude Oil	[89]
Psedomonas stutzeri *Alcaligenes denitrificans*	Musa creek, Iran	Anthracene	[93]
Isoptericola *Cellulomonas* *Bacillus* *Oceanobacillus* *Citricoccus* *Georgenia* *Marinobacter* *Micrococcus*	Kuwait, Persian Gulf	Hydrocarbon	[78, 85]
Vibrio *Planococcus* *Pseudomonas* *Actinobacterium*	Kuwait, Persian Gulf	Oil pollution	[81]

Table 1.
Bacteria removing oil pollutions separated from the Persian Gulf.

firmus was identified through biochemical tests. Examining the growth of *B. firmus* in concentrations of 50, 100, 200, 400 and 800 mg/liter showed that the maximum growth of bacteria is in the concentration of 50 mg/liter of lead. This bacterium was also able to grow well in high concentrations of lead (800 mg/L), which indicates the high resistance of the bacterium to this metal. Examining the results of the ability of *B. firmus* bacteria to bioabsorb lead at concentrations of 50, 100 and 200 mg/liter showed that this bacterium is able to remove 95% of lead from the metal solution [103].

Based on the studies of Lamouchi et al., Bacillus pastorii and *Amphibacillus xylanus* bacteria were isolated, purified and identified from the oil-contaminated sediments of Imam Khomeini Port. The mentioned bacteria grew well in culture medium containing 30 ppm anthracene. The results showed that compared to *A. xylanus*, Bacillus pastori grows and adapts quicker in the culture medium containing anthracene hydrocarbon at a temperature of 30°C [107].

A study conducted by Salamat et al., *Bacillus subtilis* resistant to anthracene and lead was isolated from the polluted sediments of the Persian Gulf. Biosurfactant production was investigated using three methods: drop collapse, blood agar and oil spreading. The maximum growth of bacteria in 30 mg/l anthracene and 50 mg/l lead was recorded by spectrophotometer. A significant proportion of anthracene (69.95%) was reduced after 120 hours, and the maximum percentage of lead absorption (82%) was observed after 150 minutes. The consequences showed that the separated bacteria are able to eliminate anthracene and lead [6].

The high capacity of the Bacillus bacterium probably owing to the species diversity in the environment and its spores which protected the bacteria from intense environmental conditions [6]. Organic compounds are considered as a nutritional source for bacteria. Specific sensors activated in the bacteria in medium comprising hydrocarbons, enabled them to bind to these organic compounds and emulsify and move them into their body [6, 106].

The results of the study by Abyar et al. showed that among different separated bacterial species (Pseudomonas, Bacillus, Alcaligenes and Geobacillus),

Bactrial name	**Isolation area**	**pollutant**	**Reference**
Pseudomonas putida strain 1389	Persian Gulf	Copper	[110]
Pseudomonas sp.	Khor Musa, Persian Gulf	Copper	[77]
Bacillus subtilis	Persian Gulf	Lead	[6]
Achromobacter denitrifican strain PQ-1	Musa Bay Persian Gulf	Cadmium	[108]
Bacillus firmus	Imam Khomeini port	Lead	[103]
Pseudomonas aeroginosa	Imam Khomeini port	Zinc	[92]
Micrococcus luteus	Persian Gulf	Zinc	[80]
Ochrobactrum anthropi	Persian Gulf	Lead	[95]
Pseudomonas aeruginosa strain MCCB 102	Persian Gulf	Copper, Zinc, Lead, Cadmium	[93]

Table 2.
Heavy metal removing bacteria isolated from sediments.

Achromobacter denitrificans strain PQ-1 was the most resistant bacteria against cadmium. Consequences also indicated that metal biosorption by *A. denitrificans* strain PQ-1 is accomplished of eliminating 55.1%, 50.7% and 43.5% of the cadmium from 25, 50, and 100 mg/l concentrations, respectively [108].

Micrococcus luteus as a zinc-resistant bacterium was isolated from Persian Gulf sediments and identified using biochemical tests and bacteriological sources. The bacteria grew well in high concentration of zinc (800 mg/liter). The results showed that the removal of zinc increases with the increase of zinc concentration, so that the highest amount of removal (169 ± 0.4 mg/L) at the concentration of 200 mg/L of zinc and the highest percentage of bioabsorption of zinc (76%) at the concentration of 50 mg/L occurred after 150 min of incubation. The results showed that *M. luteus* is resistant to changes in the percentage of salt in the environment and can grow in a wide range of salt concentration (**Tables 1** and **2**) [80].

6. Conclusion and future perspective

While environmental biotechnology is approximately based on a variety of methods of environmental protection and restoration of damaged ecosystems, for a decade it has been mostly associated with bioremediation, rehabilitation, and waste treatment technology. Most of the Persian Gulf regions, especially the coastal areas, have been contaminated with oil and heavy metals during the recent Persian Gulf War (1990) and sea transportation due to the presence of many industries. Microorganisms can degrade organic and inorganic pollutants as a source of carbon and energy, and this may pave the way for the use of remediation and cleanup methods for sediments contaminated with oil and heavy metals. There are several problems related to the environment contaminated with oil and heavy metals that affect human health and the quality of the environment. The present study may provide opportunities for biological remediation of areas contaminated with various pollutants for future studies in the Persian Gulf region.

Author details

Razieh Lamoochi[1*] and Kobra Jalali[2]

1 Department of Marine Biology, Marin Biology, Marine Science and Technology University of Khorramshahr, Iran

2 Khorramshahr Marine Science and Technology University, Iran

*Address all correspondence to: raziehlamoochi@yahoo.com

References

[1] Wang S, Xing L, Chen H. Impact of marine industrial structure on environmental efficiency. Management of Environmental Quality. 2020;**31**(1):111-129. DOI: 10.1108/MEQ-06-2019-0119

[2] Manzoor R, Zhang T, Zhang X, Wang M, Pan JF, Wang Z, et al. Single and combined metal contamination in coastal environments in China: Current status and potential ecological risk evaluation. Environmental Science Pollution Researches International. 2018;**25**:1044-1054. DOI: 10.1007/s11356-017-0526-9

[3] Hasssanshahian M, Amirnezhad N, Askarinejad Behzadi M. Crude oil pollution and biodegradation at the Persian Gulf: A comprehensive and review study. Journal of Environmental Health Science and Engineering. 2020;**18**:1415-1435. DOI: 10.1007/s40201-020-00557-x

[4] Mei G, Zhang X, Gu J. Assessment of heavy metals, polycyclic aromatic hydrocarbons, and Perfluorinated alkyl substances in two marine crustaceans (*Oratosquilla oratoria* and *Portunus trituberculatus*) in the Zhoushan fishing ground of China East Sea. Journal Ocean Uniersity China. 2021;**20**:1587-1596. DOI: 10.1007/s11802-021-4884-z

[5] Bolisetty S, Peydayesh M, Mezzenga R. Sustainable technologies for water purification from heavy metals: Review and analysis. Chemical Society Reviews. 2019;**48**(2):463-487. DOI: 10.1039/c8cs00493e

[6] Salamat N, Lamoochi R, Shahaliyan F. Metabolism and removal of anthracene and lead by a *B. subtilis*-produced biosurfactant. Toxicology Reports. 2018;**5**:1120-1123. DOI: 10.1016/j.toxrep.2018.11.004

[7] Guo W, Pan B, Sakkiah S, Yavas G, Ge W, Zou W, et al. Persistent organic pollutants in food: Contamination sources, health effects and detection methods. International Journal of Environmental Research and Public Health. 2019;**16**(22):4361. DOI: 10.3390/ijerph16224361

[8] Ali H, Khan E. Trophic transfer, bioaccumulation, and bio-magnification of non-essential hazardous heavy metals and met-alloids in food chains/webs – Concepts and implications for wildlife and human health. Human Ecological Risk Assessment. 2018;**25**:1353-1376. DOI: 10.1080/10807039.2018.1469398

[9] Filote C, Roșca M, Hlihor RM, Cozma P, Simion IM, Apostol M, et al. Sustainable application of biosorption and bioaccumulation of persistent pollutants in wastewater treatment. Current Practice Processes. 2021;**9**:1696. DOI: 10.3390/pr9101696

[10] Timková I, Sedláková-Kaduková J, Pristaš P. Biosorption and bioaccumulation abilities of Actinomycetes/Streptomycetes isolated from metal contaminated sites. Separations. 2018;**5**(4):54. DOI: 10.3390/

[11] Shao-Heng Liu, Guang-Ming Zeng, Qiu-YaNiu, Yang Liu, Lu Zhou, Lu-Hua Jiang, et al. Bioremediation mechanisms of combined pollution of PAHs and heavy metals by bacteria and fungi: A mini review. Bioresource Technology. 2017;**224**:25-33. DOI: 10.1016/j.biortech.2016.11.095

[12] Wang X, Li D, Ga P, Gu W, He X, Yang W, et al. Analysis of biosorption and biotransformation mechanism of *pseudomonas chengduensis* strain MBR under Cd (II) stress from genomic perspective. Ecotoxicology and Environmental Safety. 2020;**198**:1-9. DOI: 10.1016/j.ecoenv.2020.110655

[13] Azubuike CC, Chikere CB, Okpokwasili GC. Bioremediation techniques–classification based on site of application: Principles, advantages, limitations and prospects. World Journal Microbiology and Biotechnology. 2016;**32**:180. DOI: 10.1007/s11274-016-2137-x

[14] Abatenh E, Gizaw B, Tsegaye Z, Wassie M. The role of microorganisms in bioremediation-a review. Open Journal of Environmental Biology. 2017;**2**(1):030-046. https://www.peertechz.com

[15] Gong Y, Zhao X, Cai Z, O'Reilly SE, Hao X, Zhao D. A review of oil, dispersed oil and sediment interactions in the aquatic environment: Influence on the fate, transport and remediation of oil spills. Marine Pollution Buellton. 2014;**15**(79):16-33. DOI: 10.1016/j.marpolbul.2013.12.024 Epub 2013

[16] Ke CL, Gu YG, Liu Q, Li LD, Huang HH, Cai N. Polycyclic aromatic hydrocarbons (PAHs) in wild marine organisms from South China Sea: Occurrence, sources, and human health implications. Marine Pollution Bulletin. 2017;**117**:507-511. DOI: 10.1016/j.marpolbul.2017.02.018

[17] Honda M, Suzuki N. Toxicities of polycyclic aromatic hydrocarbons for aquatic animals. International Journal of Environmental Research and Public Health. 2020;**17**(4):1363. DOI: 10.3390/ijerph17041363

[18] Hisamuddin NH, Jalaludin J, Abu Bakar S, Latif MT. The influence of environmental polycyclic aromatic hydrocarbons (PAHs) exposure on DNA damage among school children in urban traffic area, Malaysia. International Journal Environmental Researches and Public Health. 2022;**19**:2193. DOI: 10.3390/ijerph1904219

[19] Sulong NA, Latif MT, Sahani M, Khan MF, Fadzil MF, Mohd Tahir N, et al. Distribution, sources and potential health risks of polycyclic aromatic hydrocarbons (PAHs) in PM2.5 collected during different monsoon seasons and haze episode in Kuala Lumpur. Chemosphere. 2019;**219**:1-14. DOI: 10.1016/j.chemosphere.2018.11.195

[20] Abdel-Shafy H, Mansour M. A review on polycyclic aromatic hydrocarbons: Source, environmental impact, effect on human health and remediation. Egyptian Journal of Petroleum. 2016;**25**(1):107-123. DOI: 10.1016/j.ejpe.2015.03.011

[21] Mallah MA, Changxing L, Mallah MA, Noreen S, Liu Y, Saeed M, et al. Polycyclic aromatic hydrocarbon and its effects on human health: An updated review. Chemosphere. 2022;**10**:133948. DOI: 10.1016/j.chemosphere.2022.133948

[22] Li F, Qiu ZZ, Zhang JD. Investigation, pollution mapping and simulative leakage health risk assessment for heavy metals and metalloids in groundwater from a typical brownfield, middle China. International Journal of Environmental Research and Public Health. 2017;**14**(7):768. DOI: 10.3390/ijerph14070768

[23] Krishna AK, Mohan KR. Distribution, correlation, ecological and health risk assessment of heavy metal contamination in surface soils around an industrial area, Hyderabad, India. Environment and Earth

Science. 2016;**75**:411. DOI: 10.1007/s12665-015-5151-7

[24] Sonone SS, Jadhav S, Sankhla MS, Kumar R. Water contamination by heavy metals and their toxic effect on aquaculture and human health through food chain. Letters in Applied Nano Bio Science. 2021;**10**(2):2148-2166. DOI: 10.33263/lianbs102.21482166

[25] Ali H, Khan E, Ilahi I. Environmental chemistry and ecotoxicology of hazardous heavy metals: Environmental persistence, toxicity and bioaccumulation. Journal of Chemistry. 2019;**2019**:1-14. DOI: 10.1155/2019/6730305

[26] Arantes FP, Savassi LA, Santos HB, Gomes MVT, Bazzoli N. Bioaccumulation of mercury, cadmium, zinc, chromium, and lead in muscle, liver, and spleen tissues of a large commercially valuable catfish species from Brazil. Anais Da Academia Brasileira De Ciências. 2016;**88**:137-147. DOI: 10.1590/0001-3765201620140434

[27] Jafarzadeh N, Heidari K, Meshkinian A, Kamani H, Mohammadi AA, Conti GO. Non-carcinogenic risk assessment of exposure to heavy metals in underground water resources in Saraven, Iran: Spatial distribution, Montecarlo simulation, sensitive analysis. Environmental Research. 2022;**204**:15. DOI: 10.1016/j.envres.2021.112002

[28] Ali H, Khan E. Bioaccumulation of non-essential hazardous heavy metals and metalloids in freshwater fish. Risk to human health. Environmental Chemistry Letters. 2018;**16**(3):903-917. DOI: 10.1007/s10311-018-0734-7

[29] Baki MA, Hossain MM, Akter J, Quraishi SB, Shojib MF, Ullah AA, et al. Concentration of heavy metals in seafood (fishes, shrimp, lobster and crabs) and human health assessment in Saint Martin Island, Bangladesh. Ecotoxicology and Environmental Safety. 2018;**15**(159):153-163. DOI: 10.1016/j.ecoenv.2018.04.035

[30] Gu YG, Ning JJ, Ke CL, Huang HH. Bioaccessibility and human health implications of heavy metals in different trophic level marine organisms: A case study of the South China Sea. Ecotoxicology and Environmental Safety. 2018;**15**(163):551-557. DOI: 10.1016/j.ecoenv.2018.07.114

[31] Baken KA, Sjerps RMA, Schriks M, van Wezel AP. Toxicological risk assessment and prioritization of drinking water relevant contaminants of emerging concern. Environment International. 2018;**118**:293-303. DOI: 10.1016/j.envint.2018.05.006

[32] Peña-Guzmán C, Ulloa-Sánchez S, Mora K, Helena-Bustos R, Lopez-Barrera E, Alvarez J, et al. Emerging pollutants in the urban water cycle in Latin America: A review of the current literature. Journal Environment Management. 2019;**237**:408-423. DOI: 10.1016/j.jenvman.2019.02.100

[33] Prest EI, Hammes F, van Loosdrecht MCM, Vrouwenvelder JS. Biological stability of drinking water: Controlling factors, methods, and challenges. Frontiers in Microbiology. 2016;7:45. DOI: 10.3389/fmicb.2016.00045

[34] Dueñas-Muñoz D, Guevara O, Oviedo G-R, Crisanto-Perrazo T, Toulkeridis T. Sustainable treatment techniques for emerging pollutants—The case of personal hygiene products. Applied Sciences. 2022;**12**(13):6330. DOI: 10.3390/app12136330

[35] Safiyanu I, Isah AA, Abubakar US, Rita Singh M. Review on comparative study on bioremediation for oil spills

using microbes. Research Journal of Pharmaceutical, Biological and Chemical Sciences. 2015;**6**(6):783-789. https://www.academia.edu/35935030/Review_on_Bioremediation_of_oil_spills_using_microbial_approach

[36] Azubuike CC, Chikere CB, OkpokwasiliGC.Bioremediationtechniques–classification based on site of application: Principles, advantages, limitations and prospects. World Journal of Microbiology and Biotechnology. 2016;**32**(11):1-18. DOI: 10.1007/s11274-016-2137-x

[37] Sutherland DL, Ralph PJ. Microalgal bioremediation of emerging contaminants—Opportunities and challenges. Water Research. 2019;**164**:13. DOI: 10.1016/j.watres.2019.114921

[38] Ayangbenro AS, Babalola OO. A new strategy for heavy metal polluted environments: A review of microbial biosorbents. International Journal of Environmental Research and Public Health. 2017;**14**(1):94. DOI: 10.3390/ijerph14010094

[39] Benmalek Y, Fardeau ML. Isolation and characterization of metal-resistant bacterial strain from wastewater and evaluation of its capacity in metal-ions removal using living and dry bacterial cells. International journal of environmental science and technology. 2016;**13**(9):2153-2162. DOI: 10.1007/s13762-016-1048-6

[40] Mustafa S, Bhatti HN, Maqbool M, Iqbal M. Microalgae biosorption, bioaccumulation and biodegradation efficiency for the remediation of wastewater and carbon dioxide mitigation: Prospects, challenges and opportunities. Journal of Water Process Engineering. 2021;**41**:15. DOI: 10.1016/j.jwpe.2021.102009

[41] Ungureanu CV, Favier L, Bahrim GE. Improving biodegradation of Clofibric acid by Trametes pubescens through the Design of Experimental Tools. Microorganisms. 2020;**8**(8):1243. DOI: 10.3390/microorganisms8081243

[42] Favier L, Ungureanu CV, Simion AI, Bahrim G, Vial C. Enhancing the biodegradation efficiency of a emergent refractory water pollutant by a bacterial isolate through a statistical process optimization approach. Process Safety and Environment Protection. 2021;**148**:1133-1145. DOI: 10.1016/j.psep.2021.02.012

[43] Razzak SHA, Faruque MO, Alsheikh Z, Alsheikhmohamad L, Alkuroud D, Alfayez A, et al. A comprehensive review on conventional and biological-driven heavy metals removal from industrial wastewater. Environmental Advances. 2022;**7**:26. DOI: 10.1016/j.envadv.2022.100168

[44] Sayqal A, Ahmed OB. Advances in heavy metal bioremediation: An overview. Applied Bionics and Biomechanics. 2021;**2021**:8. DOI: 10.1155/2021/1609149

[45] Acosta Rodríguez I, Cardenas-Gonzalez JF, Perez ASR, Oviedo JT, Martínez-Juarez VM. Bioremoval of different heavy metals by the resistant fungal strain aspergillus Niger. Bio Inorganic Chemistry Applied. 2018;**12**:1-13. DOI: 10.1155/2018/3457196

[46] Sultana N, Hossain SMZ, Mohammed ME, Irfan MF, Haq B, Faruque MO, et al. Experimental study and parameters optimization of microalgae based heavy metals removal process using a hybrid response surface methodology-crow search algorithm. Scientific Reports. 2020;**10**:1-15. DOI: 10.1038/s41598-020-72236-8

[47] Aslam F, Yasmin A, Sohail S. Bioaccumulation of lead, chromium, and nickel by bacteria from three different

genera isolated from industrial effluent. International Microbiology. 2020;**23**:253-261. DOI: 10.1007/s10123-019-00098-w

[48] Bowman N, Patel D, Sanchez A, Xu W, Alsaffar A, Tiquia-Arashiro SM. Lead-resistant bacteria from Saint Clair River sediments and Pb removal in aqueous solutions. Applied Microbiology and Biotechnology. 2018;**102**:2391-2398. DOI: 10.1007/s00253-018-8772-4

[49] Priyadarshanee M, Das S. Biosorption and removal of toxic heavy metals by metal tolerating bacteria for bioremediation of metal contamination: A comprehensive review. Journal of Environmental Chemical Engineering. 2021;**9**:1-14. DOI: 10.1016/j.jece.2020.104686

[50] Figuerola B, Hancock AM, Bax N, Cummings VJ, Downey R, Griffiths HJ, et al. A review and meta-analysis of potential impacts of ocean acidification on marine Calcifiers from the Southern Ocean. Frontiers in Marine Science. 2021;**8**:584445. DOI: 10.3389/fmars.2021.584445

[51] Bayani N. Ecology and environmental challenges of the Persian Gulf. Iranian Studies. 2016;**49**(6):1047-1063. DOI: 10.1080/00210862.2016.1241569

[52] Izadi A, Dobaradaran S, Nabipour I, Noroozi Karbasdehi V, Abedi E, Darabi H, et al. Data on diversity and abundance of zooplanktons along the northern part of the Persian Gulf, Iran. Data in Brief. 2018;**19**:1418-1422. DOI: 10.1016/j.dib.2018.06.012

[53] Maghsoudlou A, Momtazi F, Nasiri K. A review on the state of the biodiversity knowledge on Iran's southern seas: Introducing a methodology to evaluate the validity of the reported cases. Marine Biodiversity. 2019;**49**:563-581. DOI: 10.1007/s12526-017-0835-8

[54] Zahed MA, Rouhani F, Mohajeri S, Bateni F, Mohajeri L. An overview of Iranian mangrove ecosystems, northern part of the Persian Gulf and Oman Sea. Acta Ecologica Sinica. 2010;**30**(4):240-244. DOI: 10.1016/j.chnaes.2010.03.013

[55] Mirza R, Moeinaddini M, Pourebrahim SH, Zahed. Contamination MA: Ecological risk and source identification of metals by multivariate analysis in surface sediments of the khouran straits, the Persian Gulf. Marine Pollution Bulletin. 2019;**145**:526-535. DOI: 10.1016/j.marpolbul.2019.06.028

[56] Arfaeinia H, Dobaradaran S, Moradi M, Pasalari H, Abouee Mehrizi E, Taghizadeh F, et al. The effect of land use configurations on concentration, spatial distribution, and ecological risk of heavy metals in coastal sediments of northern part along the Persian Gulf. Science of the Total Environment. 2019;**653**:783-791. DOI: /10.1016/j.scitotenv.2018.11.009

[57] Naderi Beni A, Marriner N, Marriner N, Sharifi A, Azizpour J. Kabiri K, Djamali M, Kirman a: Climate change: A driver of future conflicts in the Persian Gulf region? Heliyon. 2021;**7**(2):1-18. DOI: 10.1016/j.heliyon.2021

[58] Hassanshahian M, Ahmadinejad M, Tebyanian H, Kariminik A. Isolation and characterization of alkane degrading bacteria from petroleum reservoir waste water in Iran (Kerman and Tehran provenances). Marine Pollution Bulletin. 2014;**73**:300-305. DOI: 10.1016/j.marpolbul.2013.05.002

[59] de Mora S, Tolosa I, Fowler SW, Villeneuve JP, Cassi R, Cattini C. Distribution of petroleum hydrocarbons and organochlorinated contaminants in marine biota and coastal sediments from the ROPME Sea area during 2005. Marine Pollution Bulletin.

2010;**60**:2323-2349. DOI: 10.1016/j.marpolbul.2010.09.021

[60] Lindén O, Jernelöv A, Egerup J. The environmental impacts of the Gulf war 1991. International Institute for Applied Systems Analysis Schlossplatz. 2004;**1**:1-94 www.iiasa.ac.at

[61] Akhbarizadeh R, Russo G, Rossi S, Golianova K, Moore F, Guida M, et al. Emerging endocrine disruptors in two edible fish from the Persian Gulf: Occurrence, congener profile, and human health risk assessment. Marine Pollution Bulletin. 2021;**166**:112441. DOI: 10.1016/j.marpolbul.2021.112241

[62] Vaezi AR, Karbassi AR, Fakhraee M. Assessing the trace metal pollution in the sediments of Mahshahr Bay, Persian gulf, via a novel pollution index. Environmental Monitoring and Assessment. 2015;**187**(10):1-12. DOI: 10.1007/s10661-015-4833-7

[63] Mehdinia A, Aghadadashi V, Sheijooni Fumani N. Origin, distribution and toxicological potential of polycyclic aromatic hydrocarbons in surface sediments from the Bushehr coast. The Persian Gulf, Marine Pollution Bulletin. 2015;**90**:334-338. DOI: 10.1016/j.marpolbul.2014.09.021

[64] Sharifinia M, Keshavarzifard M, Hosseinkhezri P, Khanjani MH, Kong Yap C, Smith WO, et al. The impact assessment of desalination plant discharges on heavy metal pollution in the coastal sediments of the Persian Gulf. Marine Pollution Bulletin. 2022;**178**:113599. DOI: 10.1016/j.marpolbul.2022.113599

[65] Ranjbar Jafarabadi A, Dashtbozorg M, Mitra S, Riyahi Bakhtiari A, Mohamadjafari Dehkordi SH, Cappello T. Historical sedimentary deposition and ecotoxicological impact of aromatic biomarkers in sediment cores from ten coral reefs of the Persian Gulf, Iran. Science of the Total Environment. 2019;**696**:133969. DOI: 10.1016/j.scitotenv.2019.133969

[66] Tasca AL, Fletcher A. State of the art of the environmental behaviour and removal techniques of the endocrine disruptor 3,4-Dichloroanilineournal. Environmental Science and Health. 2018;**53**:260-270. DOI: 10.1080/10934529.2017.1394701

[67] Xia S, Song Z, Jeyakumar P, Shaheen SM, Rinklebe J, Ok YS, et al. A critical review on bioremediation technologies for Cr (VI)-contaminated soils and wastewater. Critical Reviews in Environmental Science and Technology. 2019;**49**(12):1027-1078. DOI: 10.1080/10643389.2018.1564526

[68] Khomarbaghi Z, Shavandi M, Amoozegar MA. Bacterial community dynamics during bioremediation of alkane- and PAHs-contaminated soil of Siri Island, Persian gulf: A microcosm study. International Journal of Environmental Science and Technology. 2019;**16**:7849-7860. DOI: 10.1007/s13762-018-02198-y

[69] Herrero M, Stuckey DC. Bioaugmentation and its application in wastewater treatment: A review. Chemosphere. 2015;**140**:119-128. DOI: 10.1016/j.chemosphere.2014.10.033

[70] Alotaibi BS, Khan M, Shamim S. Unraveling the underlying heavy metal detoxification mechanisms of bacillus species. Microorganisms. 2021;**9**(8):1-30. DOI: 10.3390/microorganisms9081628

[71] Verma S, Kuila A. Bioremediation of heavy metals by microbial process. Environmental Technology & Innovation. 2019;**14**:11. DOI: 10.1016/j.eti.2019.100369

[72] Pham VHT, Kim J, Chang S, Chung W. Investigation of lipolytic-secreting bacteria from an artificially polluted soil using a modified culture method and optimization of their lipase production. Microorganism. 2021;**9**:1-13. DOI: 10.3390/microorganisms9122590

[73] Kang CH, Kwon YJ, So JS. Bioremediation of heavy metals by using bacterial mixtures. Ecological Engineering. 2016;**89**:64-69. DOI: 10.1016/j.ecoleng.2016.01.023

[74] Mekuto L, Jackson VA, Ntwampe SK. Biodegradation of free cyanide using bacillus sp. consortium dominated by bacillus Safensis, Lichenformis and Tequilensis strains: A bioprocess supported solely with whey. Journal of Bioremediation & Biodegradation. 2013;**18**:1-7. DOI: 10.4172/2155-6199.S18-00

[75] Jebelli MA, Maleki A, Amoozegar MA, Kalantar E, Gharibi F, Darvish N, et al. Isolation and identification of the native population bacteria for bioremediation of high levels of arsenic from water resources. Journal of Environmental Management. 2018;**212**:39-45. DOI: 10.1016/j.jenvman.2018.01.075

[76] Sarkar J, Kazy SK, Gupta A, Dutta A, Mohapatra B, Roy A, et al. Biostimulation of indigenous microbial Community for Bioremediation of petroleum refinery sludge. Frontiers in Microbiology. 2016;7:1407. DOI: 10.3389/fmicb.2016.01407

[77] Safahieh A, Abyar H, Roostan Z, Zolgharnein H, Mojoodi F. Isolation and characterization of pseudomonas resistant to heavy metals and poly aromatics hydrocarbons (PAHs) from Persian gulf sediments. African Journal of Biotechnology. 2012;**11**(19):4418-4423. DOI: 10.5897/AJB11.3450

[78] Al-Awadhi H, Dashti N, Kansour M, Sorkhoh N, Radwan S. Hydrocarbon-utilizing bacteria associated with biofouling materials from offshore waters of the Arabian gulf. International Biodeterioration & Biodegradation. 2012a;**69**:10-16. DOI: 10.1016/j.ibiod.2011.12.008

[79] Al-Mailem D, Eliyas M, Khanafer M, Radwan S. Culture dependent and culture-independent analysis of hydrocarbonoclastic microorganisms indigenous to hypersaline environments in Kuwait. Microbial Ecology. 2014a;**67**:857-865. DOI: 10.1007/s00248-014-0386-5

[80] Lamoochi R, Safahieh A, Salamat N, Abyar H. Study the function of halotolerant *Micrococcus luteus* in biological removal of zinc. Journal of Natural Environment. 2016;**69**(2):487-504. DOI: 10.22059/jne.2016.59760

[81] Radwan SS, Al-Hasan RH. Oil pollution and cyanobacteria, the ecology of cyanobacteria. Springer. 2000;**85**:307-319. DOI: 10.1007/0-306-46855-7_11

[82] Al-Mailem D, Eliyas M, Khanafer M, Radwan S. Culture-dependent and culture-independent analysis of hydrocarbonoclastic microorganisms indigenous to hypersaline nvironments in Kuwait. Microbial Ecology. 2014b;**67**:857-865. DOI: 10.1007/s00248-014-0386-5

[83] Al-Mailem DM, Eliyas M, Radwan S. Enhanced bioremediation of oil-polluted, hypersaline, coastal areas in Kuwait via vitamin fertilization. Environmental Science and Pollution Research. 2014c;**21**:3386-3394. DOI: 10.1007/s11356-013-2293-6

[84] Giovannoni SJ. Evolution diversity and molecular ecology of marine prokaryotes. Microbial Ecology Oceans.

2000;**12**:47-84. Available from: https://cir.nii.ac.jp/crid/1573387450712700928

[85] Al-Awadhi H, Al-Mailem D, Dashti N, Khanafer M, Radwan S. Indigenous hydrocarbon-utilizing bacterioflora in oil-polluted habitats in Kuwait, two decades after the greatest man-made oil spill. Archives Microbiology. 2012b;**194**(8):689-705. DOI: 10.1007/s11356-013-2293-6

[86] Harwati TU, Kasai Y, Kodama Y, Susilaningsih D, Watanabe K. Characterization of diverse hydrocarbon-degrading bacteria isolated from Indonesian seawater. Microbes Environment. 2007;**22**:412-415. DOI: 10.1264/jsme2.22.412

[87] Radwan S. Microbiology of oil-contaminated desert soils and coastal areas in the Arabian gulf region. In: Dion P, Nautiyal CS, editors. Microbiology of Extreme Soils. Berlin Heidelberg, Berlin, Heidelberg: Springer; 2008. pp. 275-298. DOI: 10.1007/978-3-540-74231-9_13

[88] Yakimov MM, Timmis KN, Golyshin PN. Obligate oil-degrading marine bacteria. Current Opinion in Biotechnology. 2007;**18**(3):257-266. DOI: 10.1016/j.copbio.2007.04.006

[89] Hassanshahian M, Emtiazi G, Cappello S. Isolation and characterization of crude-oil-degrading bacteria from the Persian Gulf and the Caspian Sea. Marine Pollution Bulletin. 2012;**64**(1):7-12. DOI: 10.1016/j.marpolbul.2011.11.006

[90] Parach A, Rezvani A, Assadi MM, et al. Biodegradation of heavy crude oil using Persian gulf autochthonous bacterium. International Journal Environment Researches. 2017;**11**:667-675. DOI: 10.1007/s41742-017-0059-6

[91] Kafilzadeh F, Rezaei A, Nowrooz Nejad MJ. Evaluation of phenanthrene biodegradation by indigenous bacteria isolated from mangrove sediments in the Persian Gulf. Advances in Environmental Biology. 2013;7(9):2218-2224. https://www.semanticscholar.org/paper

[92] Lamoochi R, Safahieh A, Salamat N, Abyar H. Study of P. aeroginosa capable of removing Zn (II) from aqueous solution. Environmental Sciences. 2014;**12**(1):59-66. https://envs.sbu.ac.ir/article_97385.html?lang=en

[93] Shahaliyan F, Safahieh A, Salamat N, Mojodi F, Zaredoost M. Isolation of anthracene resistant bacteria from Musa creek and scrutiny their growth and ability in biosurfactant production. Environmental Sciences. 2013;**11**(3):135-144 https://envs.sbu.ac.ir/article_97315.html?lang=en

[94] Zolgharnein H, Karami K, Mazaheri Assadi M, Dadolahi S. Investigation of heavy metals biosorption on Pseudomonas aeruginosa strain MCCB 102 isolated from the Persian Gulf. Asian Journal of Biotechnology. 2010;**2**:99-109. DOI: 10.3923/ajbkr.2010.99.109

[95] Shshalian F, Lamoochi R, Safahieh A, Salamat N, Abyar H, Mojodi F, et al. A study of the ability of bacteria isolated from Persian gulf sediments to biologically remove Lead and anthracene. Journal of Natural Environment. 2016;**69**(3):743-756. DOI: 10.22059/jne.2016.61877

[96] Varjani SJ, Upasani VN. Biodegradation of petroleum hydrocarbons by oleophilic strain of Pseudomonas aeruginosa NCIM 5514. Bioresource Technology. 2016;**222**:195-201. DOI: 10.1016/j.biortech.2016.10.006

[97] Huang Y, Zhou H, Zheng G, et al. Isolation and characterization of biosurfactant-producing Serratia marcescens ZCF25 from oil sludge

and application to bioremediation. Environmental Science and Pollution Research. 2020;**27**:27762-27772. DOI: 10.1007/s11356-020-09006-6

[98] Wasi S, Tabrez S, Ahmad M. Use of pseudomonas spp. for the bioremediation of environmental pollutants: A review. Environmental Monitoring and Assessment. 2013;**185**:8147-8155. DOI: 10.1007/s10661-013-3163-x

[99] Safdari MS, Kariminia HR, Ghobadi Nejad Z. Study potential of indigenous Pseudomonas aeruginosa and Bacillus subtilis in bioremediation of diesel-contaminated water. Water, Air, & Soil Pollution. 2017;**228**:1-7. DOI: 10.1007/s11270-016-3220-5

[100] Aryal M. Removal and recovery of nickel ions from aqueous solutions using *Bacillus sphaericus* biomass. International Journal of Environmental Research. 2015;**9**(4):1147-1156. DOI: 10.22059/ijer.2015.1004

[101] Bai J, Chao Y, Chen Y, Wang S, Qiu R. The effect of interaction between Bacillus subtilis DBM and soil minerals on Cu (II) and Pb (II) adsorption. Journal of Environmental Sciences. 2019;**78**:328-337. DOI: 10.1016/j.jes.2018.11.012

[102] Pi H, Helmann JD. Sequential induction of fur-regulated genes in response to iron limitation in Bacillus subtilis. Proceedings of the National Academy of Sciences. 2017;**114**(48):12785-12790. DOI: 10.1073/pnas.1713008114

[103] Safahieh A, Lamoochi R, Salamat N, Abyar H. Isolation and identification of *Bacillus firmus* from the marine sediments of imam Khomeini port and study its ability in biosorption of Lead. Journal of Oceanography. 2014;**5**(17):11-19 http://joc.inio.ac.ir/article-1-515-en.html

[104] Zeng W, Li F, Wu C. Role of extracellular polymeric substance (EPS) in toxicity response of soil bacteria bacillus sp. S3 to multiple heavy metals. Bioprocess and Biosystems Engineering. 2020;**43**:153-167. DOI: 10.1007/s00449-019-02213-7

[105] Afzal MJ, Khan MI, Cheema SA. Combined application of bacillus sp. MN-54 and phosphorus improved growth and reduced lead uptake by maize in the lead-contaminated soil. Environmental Science and Pollution Research. 2020;**27**:44528-44539. DOI: 10.1007/s11356-020-10372-4

[106] Xing SC, Chen JY, Lv N, Mi JD, Chen WL, Liang JB, et al. Biosorption of lead (Pb2+) by the vegetative and decay cells and spores of *Bacillus coagulans* R11 isolated from lead mine soil. Chemosphere. 2018;**211**:804-816. DOI: 10.1016/j.chemosphere.2018.08.005

[107] Lamoochi R. Shahaliyan F, Safahieh a, Salamat N, Mojodi F, Abyar H: Isolation of anthracene resistant positive gram bacteria from imam Khomeini port oil polluted sediments. Journal of Biosafety. 2019;**11**(4):1-14 http://journalofbiosafety.ir/article-1-85-en.html

[108] Abyar H, Safahieh A, Zolgharnein H, Zamani I. Isolation and identification of Achromobacter denitrificans and evaluation of its capacity in cadmium removal. Polish Journal of Environmental Studies. 2012;**21**(6):1523-1527 http://www.pjoes.com/Isolation-and-Identification-of-Achromobacter-r-ndenitrificans-and-Evaluation-of,88898,0,2.html

[109] Al-Mailem DM, Al-Deieg M, Eliyas M, Radwan SS. Biostimulation of indigenous microorganisms for bioremediation of oily hypersaline microcosms from the Arabian gulf

Kuwaiti coasts. Journal of Environmental Management. 2017;**193**:576-583. DOI: 10.1016/j.jenvman.2017.02.054

[110] Abyar H, Mojodi F, Safahieh A, Zolgharnein H, Zamani I. The role of pseudomonas putida in bioremediation of naphthalene and copper. World Journal of Fish and Marine Sciences. 2011;**5**:444-449

Chapter 8

Bacterial Assisted Phytoremediation

Igbonomi Emmanuel Sunday and Ajayi Ochechevesho Joan

Abstract

Bacterial assisted phytoremediation describes how bacteria, particularly those found in the rhizosphere, can assist plants known as hyperaccumulators in removing heavy metal contamination from the environment. The function of hyperaccumulation is dependent not only on the plant, but also on the interaction of plant roots with rhizosphere microbes and soil bioavailable metal concentrations. Bioremediation is the process of utilizing microorganisms, fungi, green plants, or their enzymes to repair the natural environment that has been harmed by contaminants to its original state. The best plant species for phytoremediation should be hardy, produce a lot of biomass, be resistant to the toxic effects of metals and contaminants, be unappealing to herbivores. Plant growth-promoting bacteria can encourage soil fertility and health, improve plant diseases. They promote the phytoremediation process either by reducing the toxicity of pollutants or increasing the availability of pollutants or promoting the growth of plants. Metal ions bind to the cell wall's functional groups (amine, carboxyl, hydroxyl, phosphate, sulfate, amine). The effects of stresses, which are brought on whenever changes in metabolism occur, are avoided by plants using a variety of tolerance mechanims and pathways called phytohormone. Heavy metal phytoextraction involves the following steps: Intake of heavy metals by plant roots, translocation of heavy metal ions from roots to aerial parts of plants, and sequestration and compartmentation of heavy metal ions in plant tissue. As bacterial siderophores aid in reducing the stress caused by metal contaminants. Rhizosphere acidification is a common mechanism used by plant with rhizosphere's bacteria in dealing with low Phosphorus stress, to activate and increase the efficiency of soil Phosphorus utilization. The interaction between bacteria and plants has been found to be helpful in handling various pollutants in various exosystems.

Keywords: hyperaccumulator, rhizosphere, chelating agents, phytoremediation, bacteria

1. Introduction

Heavy metal contamination of soil has resulted from anthropogenic activities such as industrialization and agricultural practices. Heavy metals are typically defined as metals with densities greater than 5 g/cm^3 [1]. Metals are classified as essential or nonessential. Essential metals are metals that are required for normal cellular growth, such as zinc, nickel, copper, and so on. These heavy metals are required in low concentrations (nM), but at higher concentrations (M to mM), they are harmful to organisms [2]. Nonessential metals are metals that have no known biological function, such as lead, cadmium, and mercury [3]. At any concentration, these metals are toxic. Heavy metals have severe toxic effects on plants, animals, and human health, making remediation critical. In this context, insitu stabilization has been regarded as an effective method of remediating metal contaminated soil [4]. Among the various techniques used, phytoremediation is regarded as one of the safest, most innovative, and effective tools for heavy metal remediation. Nonessential metals include, for example, lead, cadmium, and mercurypollution has increased dramatically as industries such as mining have expanded, particularly in developing country cities. Different approaches to removing these toxic contaminants from our environment are being considered. This situation could be resolved and accelerated by investigating the plant bacteria partnership, which would improve plant growth in toxic heavy metal polluted environments and, ultimately, improve phytoremediation efficiency [5]. Bacterial assisted phytoremediation describes how bacteria, particularly those found in the rhizosphere, can assist plants known as hyperaccumulators in removing heavy metal contamination from the environment. The function of hyperaccumulation is dependent not only on the plant, but also on the interaction of plant roots with rhizosphere microbes and soil bioavailable metal concentrations [6]. Bacteria aided Phytoremediation is a lowcost and environmentally friendly method of combating heavy metal contamination in soil. Using a plant and beneficial bacteria to remediate heavy metals in soil [7]. Beneficial bacteria help plants absorb metals by mobilizing heavy metals in the soil and increasing plant vigor. Nonetheless, in unsterilized soil, effective interaction between plant and bacteria is not guaranteed, affecting phytoremediation efficiency [7]. Beneficial bacteria help plants absorb metals by mobilizing heavy metals in the soil and increasing plant vigor. Nonetheless, in unsterilized soil, effective interaction between plant and bacteria is not guaranteed, affecting phytoremediation efficiency [8, 9].

Plant-associated bacteria have the potential to promote plant growth and resistance to stress. By controlling growth hormone, ensuring nutrition security, producing siderophore and secondary metabolites, and improving the antioxidant enzyme system, these bacteria have the potential to improve plant growth. Bacteria assisted phytoremediation is a widely used integrated technology that uses bacteria and plants to degrade contaminants while having the least possible impact on the environment.

It is an effective method for cleaning up polluted soil. Sites contaminated with heavy metals, hydrocarbons, and pesticides are examples. Endophytic bacteria and rhizospheric bacteria are two types of bacteria that aid in phytoremediation. Plants and bacteria are both biological agents for heavy metal contamination control [10]. The process of removing heavy metals from soil using only bacteria is known as biosorption, whereas the process of removing heavy metals using plants is known as phytoremediation. These two processes will be discussed in more detail shortly.

1.1 Bioremediation

Bioremediation is a potential and prospective method for the treatment of contaminated environments using microbes and plants for biodegradation or biotransformation of contaminants. Chemical and physical methods of remediation are not environmentally sustainable. Bioremediation, in contrast to these methods, relies on biological processes that are mediated by various groups of living organisms to achieve the permanent removal of these pollutants [11]. The process known as "bioremediation" involves using microorganisms to reduce the amount of hazardous environmental waste that is present at a contaminated site [12]. Bioremediation is the process of utilizing microorganisms, fungi, green plants, or their enzymes to repair the natural environment that has been harmed by contaminants to its original state. Heavy metal bioremediation is the process of removing heavy metals from soil and waste water through metabolically mediated or physico-chemical pathways. The biological potentials of algae, bacteria, fungi, and yeasts are demonstrating their efficacy in the removal of metal from waste waters [13]. With the ability to degrade contaminants using natural microbial activity by various consortia of microbial strains, bioremediation is a very good alternative to conventional methods for removing heavy metal from the environment and media [14]. The most common biosorbents are living and dead microbial cells of bacteria, algae, and fungi. In this context, we'll concentrate on one type of microorganism, bacteria. The bacterial cell wall is the first component to come into contact with the metal ion during biosorption. Metal ions bind to the cell wall's functional groups (amine, carboxyl, hydroxyl, phosphate, sulfate, amine) [15]. The general metal uptake process involves metal ions binding to reactive groups on the bacterial cell wall, followed by metal ion incorporation inside the cell [16]. Bacteria have evolved a number of efficient systems for detoxifying metal ions; these resistance mechanisms are developed primarily for survival. Several living and dead bacterial strains have been studied and found to be effective at heavy metal biosorption. Several physiochemical parameters, including pH, temperature, biomass dosage, initial metal concentration, contact time, heavy metal type, and biosorbent type, have been reported to influence heavy metal biosorption by bacterial biomass processes. The functional groups that are present on the surface of the biosorbent and the structure of proteins can both be influenced by temperature. PH has an impact on the active binding sites and solubility of heavy metal ions in biomass. Numerous investigations have been made into the biosorption capacities and efficacies of different bacterial strains, as well as the influence of the process's influencing factors. According to the data that has been published, there is not a clear focus on the biosorption of heavy metals by bacterial strains that could reveal the best bacterial biosorbent and ideal physio-chemical conditions for the removal of each heavy metal.

1.2 Phytoremediation

Plants, which is both economically advantageous and environmentally beneficial [17]. In order to remove or degrade the pollutants, phytoremediation employs a variety of mechanisms, such as degradation (phytodegradation, rhizodegradation), accumulation (phytoextraction, rhizofiltration), dissipation (phytovolatilization), and immobilization (hydraulic control and phytostabilization). Plants use one or more of these mechanisms to lower the concentrations of contaminants in soil and water, depending on the contaminants. For instance, plants can absorb and store heavy metals in their tissues and degrade organic pollutants to lessen their toxicity in soil and water resources [18]. Depending on the types of contaminants and media, different plants use various techniques or combinations of them to remediate soil and water. Rhizofiltration, phytodegradation, phytovolatilization, rhizodegradation, and phytodegradation are all methods for cleaning up contaminated ground water. Rhizofiltration, phytodegradation, or rhizodegradation are three treatment options for surface and wastewater contamination. Surface and wastewater contaminations can be treated by rhizofiltration, phytodegradation or rhizodegradation. Through phytoextraction, phytodegradations, phytostabilization, rhizodegradation, or phytovolatilization, contamination caused by soil, sediments, or sludge is remedied. The best plant species for phytoremediation should be hardy, produce a lot of biomass, be resistant to the toxic effects of metals and contaminants, be simple to grow, have a high absorption capacity, and be unappealing to herbivores [19]. The harmful heavy metals in the contaminated ecosystem can be detoxified and accumulated in the plant using the efficient, environmentally friendly, and cost-effective bioremediation technique known as phytoremediation. The heavy metals present in the soil are transported and translocated to various plant parts by molecules known as transporters that hyperaccumulators exude. Higher concentrations of harmful heavy metals can be contained in the tissues of plants with hyperaccumulator genes. The effectiveness of phytoremediation depends on a number of factors, including the nature of the rhizosphere, the characteristics of the rhizosphere microflora, the soil properties (pH and soil type), organic matter in the soil, the type of heavy metal, and more. Phytoremediation approach using metal accumulating plants is much convincing in terms of metal removal efficiency, but it has many limitations because of slow plant growth and decreased biomass owing to metal-induced stress. In addition, constrain of metal bioavailability in soils is the prime factor to restrict its applicability.

1.3 Bacteria interactions with plant in detoxifying heavy metal pollutions

The current review covers in great detail the environmental effects of heavy metal toxicity as well as various phytoremediation mechanisms for the transport and accumulation of heavy metals from polluted soil. The assistance of Plant Growth Promoting (PGP) bacteria increases the effectiveness of phytoremediation (i.e rhizobacteria). These bacteria can be found in the soil near plant roots, where they use siderophores, organic acids, biosurfactants, biomethylation, and redox processes to convert toxic metals into bioavailable and soluble forms. Additionally, PGP bacteria produce phytohormones, ACC (1-aminocyclopropane-1-carboxylic acid) deaminase, phosphorus solubilization, nitrogen fixation, iron sequestration, and other growth promoting traits that help plants grow and produce more biomass, aiding in phytoremediation. In many cases, plant species cannot perform well and need aids to enhance phytoremediation. Such aids include soil amendments like biochar [20]. **Table 1** shows the important microorganism in bacteria interaction with plant. Ethylene Diamine Tetraacetic Acid (EDTA) [31], endophytic bacteria, Arbuscular mycorrhiza or even transgenic plants [32]. The bacterial rhizosphere population will directly induce root growth and thereby stimulate plant growth, improve metal resistance, and plant health [26, 33]. Plant growth-promotingbacteria have been shown to get a wide ability to increase phytoremediation efficiency. Plant growth-promoting bacteria can encourage soil fertility and health, improve plant diseases, enhance crop resistance to toxic metals, enhance plant nutrient uptake and metal absorption, and efflux. They promote the phytoremediation process either by reducing the toxicity of pollutants or increasing the availability of pollutants or promoting the growth of plants. Factors affecting this processes aids or stimulate or are either specific to soil or water resources. Comprehensive information on the contribution of each phytoremediation mechanism is showned in **Table 2**. Immunization through plant growth encouraging bacteria to produce growth factors, trigger phytopathogenic resistance, and balance plant hormones and nutrition. Other mechanisms include nutrient mobilization, exopolysaccharide development, and rhizobiotine stimulation of plant growth.

Pseudomonas libanensis, Pseudomonas aeruginosa. Brevibacterium frigoritolerans and *Bacillus paralicheniformis Bacillus cereus Brucella* sp. *Xanthomonas oryzae, Rhizospus oryzae Microbactetium* sp.	*Helianthus annuus Lens esculenta L. esculenta Hibiscus esculentus* L. *Oryza sativa* Pisium sativum	production of siderophone, Phosphate solubilization, Phytoremediation of metals in association with phosphate-solubilizing bacteria (PSB) considerably overcomes the practical drawbacks imposed by metal stress on plants Mobilization of cadmium (Cd) and lead (Pb) in soil. Improve stem and root length and weight of grain number Enhancement of plant growth, involved in increased uptake of nitrogen, synthesis of phytohormones, solubilization of minerals such as phosphorus, and secretion of siderophores that chelate iron and make it available to the plant root for improved stem and root length Fortifies plant and decreases plant contamination with Cr(VI), toxic metal species are accumulated in plant tissues through phytostabilization in extremely polluted sites High resistance to disease, mechanism are regulated by gene. Improves the plant high adsorption capacity and high effectivity to adsorb the metal ions through a process of ion exchange. Decreases Cr(IV) toxicity. Supply phytohormones and biocontrol the uptake and immobilization of chromium in soil	[21–23] [24, 25] [2, 10, 26, 27] [17, 18, 28] [1, 10, 21, 22] [28–30]

Table 1.
Some important bacteria interaction with plant and their importance.

1.4 Some important mechanisms used in bacterial assisted phytoremediation

1.4.1 Biosorption

The removal of metal or metalloid species, compounds, and particulates from a solution by low-cost biological materials is known as biosorption [40]. Biosorption is a physiochemical process that removes heavy metals from contaminated water through ion exchange, surface complexation, chelation, and metal ion coordination using biological adsorbents [41]. In general, biosorption is regarded as an efficient, lowcost, and environmentally friendly remediation method for heavy metal removal from contaminated water. The most common biosorbents are living and dead microbial cells of bacteria, algae, and fungi. In this context, we'll concentrate on one type of microorganism, bacteria. The bacterial cell wall is the first component to come into contact with the metal ion during biosorption. Metal ions bind to the cell wall's functional groups (amine, carboxyl, hydroxyl, phosphate, sulfate, amine) [15]. The general metal uptake process involves metal ions binding to reactive groups on the bacterial cell wall, followed by metal ion incorporation inside the cell [16]. Bacteria have evolved a number of efficient systems for detoxifying metal ions; these resistance mechanisms are developed primarily for survival. Several living and dead bacterial strains have been studied and found to be effective at heavy metal biosorption. Several physiochemical parameters, including pH, temperature, biomass dosage, initial metal concentration, contact time, heavy metal type, and biosorbent type, have been reported to influence heavy metal biosorption by bacterial biomass processes. Temperature can change the structure of proteins and the functional groups available on the biosorbent's surface. The solubility

Phytovolatilization	Toxicants are Lost to the atmosphere through evapotranspiration from dermal layers of stems and leaves	[29, 32]
Rhizodegradation	The metabolism of toxicants by root enzymes	[13, 32, 34]
Phytodesalination	Uptake of mineral ions	[22, 32, 35]
Phytostabilization	Formation of residue bound molecules and complexes with biomolecules	[2, 36]
Phytoextraction	Metal uptake and formation of ligands and chelation	[21, 28, 37]
Rhizofiltratiom	Uptake from ground water followed by precipitation inside root	[28, 32]
Phytodegradation	metabolism of toxicants by phyto-enzymes	[31, 38]
Rhizobial and soil living microorganisms	Help in mineralization and plant uptake of toxicants	[30, 32, 39]

Table 2.
Some phytoremediation mechanism.

of heavy metal ions as well as active binding sites in biomass are affected by PH. Several studies have been conducted to investigate the biosorption efficiencies and capacities of various bacterial strains, as well as the impact of the above-mentioned influential parameters on the process. On the basis of published data, there is not a clear focus on the biosorption of heavy metals by bacterial strains that could reveal the most effective bacterial biosorbent and ideal physio-chemical conditions for the removal of each heavy metal.

1.4.2 Phytohormones

Abiotic stresses that affect plants include salinity, heavy metals, drought, and extremely high or low temperatures. Abiotic stresses negatively affect a plant's physiology and morphology due to errors in the genetic control of cellular pathways. The effects of stresses, which are brought on whenever changes in metabolism occur, are avoided by plants using a variety of tolerance mechanisms and pathways. Phytohormones are some of the most significant growth regulators; they are well known for having a significant impact on plant metabolism and also play a significant role in the stimulation of plant defense response mechanisms against stresses. Supplementing with exogenous phytohormones has been used to enhance growth and metabolism in stressful environments [24]. Recent studies have demonstrated that phytohormones made by microbes associated with roots may prove to be significant metabolic engineering targets for promoting host tolerance to abiotic stresses. Several genetic and biochemical techniques have been used to pinpoint the biosynthetic pathways for phytohormones methods.

1.4.3 Phytoextraction

Phytoextraction is the use of plants to translocate and accumulate contaminants in their above ground biomass by absorbing them from soil or water [28, 37]. The most significant phytoremediation method used today to remove heavy metals and metalloids from contaminated soil is phytoextraction [21]. Phytoextraction is a long term method for removing heavy metals from contaminated soil, in contrast to phytostabilization, in which plants only momentarily contain heavy metals that still remain belowground. It is therefore better suited for commercial use. Heavy metal phytoextraction involves the following steps: Intake of heavy metals by plant roots, mobilization of heavy metals in the rhizosphere, translocation of heavy metal ions from roots to aerial parts of plants, and sequestration and compartmentation of heavy metal ions in plant tissues are all examples of heavy metal ion transport in plants. [21]. The efficiency of phytoextraction relies on a few factors such as plant selection, plant performance, heavy metal bioavailability, soil, and rhizosphere properties.

1.4.4 Siderophore secretion

Metal ions are bound by siderophore secretion, which reduces metal bioavailability. Low molecular weight siderophores, which are iron chelators with an exceptionally strong affinity for ferric iron (Fe3+), are secreted by bacteria in iron- limited conditions in order to acquire iron [10]. They can chelate a variety of other metals, including magnesium, manganese, chromium (III), gallium (III), cadmium, zinc, copper, nickel, arsenic, and lead, as well as radionuclides like plutonium (IV), with varying affinities, despite having a preference for Fe3+ [1, 6, 21]. As bacterial siderophores aid in reducing the stress caused by metal contaminants, the supply of iron to growing plants under heavy metal pollution becomes more important.

1.4.5 Rhizosphere acidification

Rhizosphere acidification, a common mechanism by plant with help from rhizosphere's bacteria to deal with low Phosphorus stress, to activate and increase the efficiency of soil Phosphorus utilization. Due to the release of organic acids and H+, white lupin (Lupinus albus L.) has evolved what are known as proteoid roots to acidify infertile soil when Phosphorus deficiency exists [22, 42–45]. To solubilize soil Phosphorus, sugar beet roots primarily exude salicylic acid and citramalic acid [5]. Under low-Phosphorus conditions, root-secreted acid phosphatase activity was increased in a Phosphorus efficient genotype of rapeseed (Brassica napus L.), which was advantageous for the increased uptake of Phosphorus [30].

2. Conclusion

Environmental technique that holds promise for purging the environment of heavy metal contamination is bacterial assisted phytoremediation of heavy metals. It is a safe environmental remediation technique. The phytoremediation mechanism from plants can be used by consortia of potential plant growth promoting rhizobacteria, degrading bacteria, as well as endophytic bacteria to reduce heavy metals in soil. The interaction between bacteria and plants has been found to be helpful in handling various pollutants in various exosystems.

Author details

Igbonomi Emmanuel Sunday* and Ajayi Ochechevesho Joan
University of Benin, Nigeria

*Address all correspondence to: igbonomiemma@gmail.com

References

[1] Nies DH. Microbial heavy-metal resistance. Applied Microbiology and Biotechnology. 1999;**51**:730-750

[2] Grosse C, Anton A, Hoffmann T, Franke S, Schleuder G, Nies DH. Identification of a regulatory pathway that controls the heavy-metal resistance system Czc via promoter czcNp in Ralstonia metallidurans. Archives of Microbiology. 2004;**182**(2-3):109-118

[3] Kang SH, Singh S, Kim JY, Lee W, Mulchandani A, Chen W. Bacteria metabolically engineered for enhanced phytochelatin production and cadmium accumulation. Applied and Environmental Microbiology. 2007;**73**(19):6317-6320

[4] Janssen PJ, Houdt RV, Moors H, Monsieurs P, Morin N, Michaux A, et al. The complete genome sequence of Cupriavidus metallidurans strain CH34, a master survivalist in harsh and anthropogenic environments. PLoS One. 2010;**5**(5):1-33

[5] Kumar V, Chandra R. Bacteria-assisted phytoremediation of industrial waste pollutants and ecorestoration. Phytoremediation of Environmental Pollutants. 2017;**1**:159-200

[6] Rajkumar M, Prasad MNV, Freitas H, Noriharu AE. Biotechnological applications of serpentine soil bacteria for phytoremediation of trace metals. Critical Reviews in Biotechnology. 2009;**29**(2):120-130

[7] Bell TH, Joly S, Pitre FE, Yergeau E. Increasing phytoremediation efficiency and reliability using novel omics approaches. Trends in Biotechnology. 2014;**32**(1):271-280

[8] Alka S, Shahir S, Ibrahim N, et al. The role of plant growth promoting bacteria on arsenic removal: A review of existing perspectives. Environmental Technology Innovation. 2020;**17**:100-620

[9] Chuaphasuk C, Prapagdee B. Effects of biochar-immobilized bacteria on phytoremediation of cadmium-polluted soil. Environmental Science Pollution Resources. 2019;**26**(1):23679-23688

[10] Sharma RR, Singh D, Singh R. Biological control of postharvest diseases of fruits and vegetables by microbial antagonists: A review. Biological Control. 2009;**50**(3):205-221

[11] Azubuike CC, Chikere CB, Okpokwasili GC. Bioremediation: an eco-friendly sustainable technology for environmental management. Bioremediation of Industrial Waste for Environmental Safety. 2020;**1**(1):19-39

[12] Boopathy R. Factors limiting bioremediation technologies. Bioresource Technology. 2000;**74**(1):63-67

[13] Vergani L, Mapelli F, Zanardini E, Terzaghi E, Di Guardo A, Morosini C, et al. Phyto-rhizoremediation of polychlorinated biphenyl contaminated soils: An outlook on plant-microbe beneficial interactions. Science Total Environment. 2017;**575**(1):1395-1406

[14] Juwarkar AA, Singh SK, Mudhoo AA. comprehensive overview of elements in bioremediation. Reviews in Environmental Science and Biotechnology. 2010;**9**(3):215-288

[15] Bilal M, Rasheed T, Sosa-Hernandez JE, Raza A, Nabeel F, Iqbal HMN. Biosorption: An interplay between marine algae and potentially toxic elements-a review. Marine

Compounds Used in Biosorption. 2018;**16**(2):65

[16] Abbas SH, Ismail IM, Mostafa TM, Sulaymon AH. Biosorption of heavy metals: A review. Journal of Chemical Science and Technology. 2014;**3**(4):74-102

[17] Ashraf S, Ali Q, Ahmad Z, Ashraf S, Naeem Asghar H. Phytoremediation: Environmentally sustainable way for reclamation of heavy metal polluted soils. Ecotoxicology and Environmental Safety. 2019;**174**(1):714-727

[18] Mahar A, Wang P, Ali A, Awasthi MK, Lahori AH, Wang Q, et al. Challenges and opportunities in the phytoremediation of heavy metals contaminated soils: A review. Ecotoxicology and Environmental Safety. 2016;**126**(1):111-121

[19] Adesodun JK, Atayese MO, Agbaje TA, Osadiaye BO, Mafe OF, Soretire AA. Phytoremediation potentials of sunflowers (Tithonia diversifolia and Helianthus annuus) for metals in soils contaminated with zinc and lead nitrates. Water, Air, and Soil Pollution. 2010;**207**(1):195-201

[20] Paz-Ferreiro J, Lu H, Fu S, Mendez A, Gasco G. Use of phytoremediation and biochar to remediate heavy metal polluted soils: A review. Solid Earth. 2014;**5**(1):65-75

[21] Ali H, Khan E, Sajad MA. Phytoremediation of heavy metals-concepts and applications. Chemosphere. 2013;**91**(7):869-881

[22] Li M, Osaki M, Rao IM, Tadano T. Secretion of phytase from the roots of several plant species under phosphorus-deficient conditions. Plant and Soil. 1997;**195**(1):161-169

[23] Ma Y, Rajkumar M, Oliveira RS, Zhang C, Freitas H. Potential of plant beneficial bacteria and arbuscular mycorrhizal fungi in phytoremediation of metal contaminated saline soils. Journal of Hazard Materials. 2019;**379**(2019):120-813

[24] Faisal M. Inoculation of plant growth promoting bacteria Ochrobactrum intermedium, Brevibacterium sp. and Bacillus cereus induce plant growth parameters. Journal of applied. Biotechnology. 2013;**1**(1):45-53

[25] Sheng XF, Xia JJ, Jiang CY, He LY, Qian M. Characterization of heavy metal-resistant endophytic bacteria from rape (Brassica napus) roots and their potential in promoting the growth and lead accumulation of rape. Environmental Pollution. 2008;**156**:1164-1170

[26] Fasani E, Manara A, Martini F, Furini A, DalCorso G. The potential of genetic engineering of plants for the remediation of soils contaminated with heavy metals. Plant Cell Environment. 2018;**41**(5):1201-1232

[27] Kuriakose SV, Prasad MNV. Cadmium stress affects seed germination and seedling growth in Sorghum bicolor (L.) Moench by changing the activities of hydrolyzing enzymes. Plant Growth Regulation. 2008;**54**(2):143-156

[28] Wani PA, Khan MS, Zaidi A. Chromium-reducing and plant growth-promoting Mesorhizobium improves chickpea growth in chromium-amended soil. Biotechnology Letters. 2008;**30**(1):159-163

[29] Soni SK, Singh R, Singh M, Awasthi A, Wasnik K, Kalra A. Pretreatment of Cr (VI)-amended soil with chromate-reducing rhizobacteria decreases plant toxicity and increases the yield of Pisum sativum. Archives of Environmental Contamination and Toxicology. 2014;**66**(4):616-627

[30] Zhang F, Shen J, Zhang J, Zuo Y, Li L, Chen X. Rhizosphere processes and management for improving nutrient use efficiency and crop productivity: Implications for China. Advances in Agronomy. 2010;**107**:1-32

[31] Shahid M, Pourrut B, Dumat C, Nadeem M, Aslam M, Pinelli E. Heavy-metal-induced reactive oxygen species: Phytotoxicity and physicochemical changes in plants. Reviews of Environmental Contamination and Toxicology. 2014;**232**(1):1-44

[32] Afzal M, Khan QM, Angela S. Endophytic bacteria: Prospects and applications for the phytoremediation of organic pollutants. Chemosphere. 2014;**117**(1):232-242

[33] Gupta DK, Huang HG, Corpas FJ. Lead tolerance in plants: Strategies for phytoremediation. Environmental Science Polluion Resources. 2013;**20**(4):2150-2161

[34] Hou J, Liu W, Wang B, Wang Q, Luo Y, Franks AE. PGPR enhanced phytoremediation of petroleum contaminated soil and rhizosphere microbial community response. Chemosphere. 2015;**138**:592-598

[35] Rabhi M, Ferchichi S, Jouini J, Hamrouni MH, Koyro HW, Ranieri A, et al. Phytodesalination of a salt-affected soil with the halophyte Sesuvium portulacastrum L. to arrange in advance the requirements for the successful growth of a glycophytic crop. Bioresource Technology. 2010;**101**(17):6822-6828

[36] Bolan NS, Park JH, Robinson B, Naidu R, Huh KY. Phytostabilization: A green approach to contaminant containment. Advances in Agronomy. 2011;**112**:145-204

[37] Verret F, Gravot A, Auroy P, Leonhardt N, David P, Nussaume L, et al. Overexpression of AtHMA4 enhances root-to-shoot translocation of zinc and cadmium and plant metal tolerance. FEBS Letters. 2004;**576**(3):306-312

[38] Das PK. Phytoremediation and Nanoremediation: Emerging techniques for treatment of acid mine drainage water. Defence Life Science Journal. 2018;**3**(2):190-196

[39] Diab A, Badry KA. Biodegradation of PAH compounds in the rhizosphere of Tamarix nilotica: A salt tolerant wild plant. Journal of American Science. 2011;**7**(6):115-124

[40] Wang J, Chen C. Biosorbents for heavy metals removal and their future. Biotechnology Advances. 2009;**27**(2):195-226

[41] Michalak I, Chojnacka K, Witek-krowiak A. State of the art for the biosorption process-a review. Applied Biochemistry and Biotechnology. 2013;**170**(1):1389-1416

[42] Watt M, Evans JR. Linking development and determinacy with organic acid efflux from proteoid roots of white lupin grown with low phosphorus and ambient or elevated atmospheric CO_2 concentration. Plant Physiology. 1999;**120**(3):705-716

[43] Abdi O, Kazemi MA. Review study of biosorption of heavy metals and comparison between different biosorbents. Journal of Materials and Environmental Science. 2015;**6**(5):1386-1399

[44] Ramrakhiani L, Ghosh S, Sarkar S, Majumdar S. Heavy metal biosorption in multi component system on dried activated sludge: Investigation of adsorption mechanism by surface characterization. Materials Today: Proceedings. 2016;**3**(10):3538-3552

[45] Thakur S, Singh L, Ab Wahid Z, Siddiqui MF, Atnaw SM, Din MFM. Plant-driven removal of heavy metals from soil: Uptake, translocation, tolerance mechanism, challenges, and future perspectives. Environmental Monitoring and Assessment. 2016;**188**(4):206

Chapter 9

Perspective Chapter: Can Natural Attenuation be Considered as an Effective Solution for Soil Remediation?

Clarisse Balland-Bolou-Bi, Florence Brondeau and My Dung Jusselme

Abstract

Natural attenuation is described as a naturally occurring process, mostly in soils and also in groundwater, without human intervention, which transforms, reduces and destroys the organic and inorganic contaminants. As an eco-friendly, cost-effective and relatively simple technology, natural attenuation is widely used for the treatment of contaminated soils. However, the application of this technology must be carefully controlled and monitored not only for its efficiency and durability over time, but also for the migration of contaminants to ensure no risk to human health and ecosystems. Furthermore, the success of this technique requires a good knowledge of the type of contaminants, the physical and chemical characteristics of the soils, as well as the living actors, including plants, fauna, microorganisms and their interactions, that live in the soils to be treated and that will be involved in this process. The purpose of this chapter is to provide the most recent information regarding the principle of this technology, the role of the living actors and the interactions between plant, fauna and microorganisms, the advantages and disadvantages, and finally to discuss the efficiency of this technique in comparison with other techniques such as phytoremediation or bioremediation. In fine, we will discuss its social acceptability.

Keywords: natural attenuation, soil remediation, social acceptation, organic and inorganic pollutants, plant-fauna-microorganisms interaction

1. Introduction

Several hundred thousand industrial and agricultural sites are polluted in the world today due to an increase in the production, extraction, use, and disposal of chemicals since the industrial and green revolutions. All these anthropic activities, along with natural sources of contaminants, release potentially toxic elements and compounds into the environment that have different behaviors depending on soil properties and their interactions with the ecosystem. The main reasons that contaminants are found in the critical zone are numerous: (i) extraction of minerals and raw materials; (ii) use of agrochemicals, untreated organic wastes as sources of fertilizers and irrigating with wastewater; (iii) insufficient management of urban and industrial and hazardous waste; (iv) combustion of raw materials for energy production; and (v) development of widespread transport networks and infrastructures [1]. As a consequence, it is important to know that nearly half of the soils in the world are degraded [2]. The remediation of these sites becomes then almost impossible in spite of the more or less advanced referencing of the administrations throughout the world. Thus, several billion square meters of polluted land were left abandoned without human intervention. Over time, in these soils, a natural phenomenon of soil remediation, called natural attenuation, will be developed in which all live microorganisms that interact closely with the environmental parameters play an important role in this. Today, natural attenuation shows its good capacity to remediate polluted soils. However, numerous questions arise when applying this technology (i) how these soils can manage the pollution?; (ii) what is the evolution of the remediation system in these polluted soils?; (iii) how nature reclaimed its rightful place, and (iv) how is the acceptance of this technique at different levels: science, politics,s and social.

2. An overview of natural attenuation

According to the United States Environmental Protection Agency [3], the natural attenuation (NA) processes include a variety of processes that reduce the mass, toxicity, mobility, volume, or concentration of contaminants in soil and groundwater. Natural attenuation takes place without human intervention. The *in situ* processes can refer to physical, chemical, or biological processes and include biodegradation, dispersion, dilution, sorption, volatilization, radioactive decay, chemical or biological stabilization, transformation, and destruction of contaminants. Monitored natural attenuation (MNA) refers to the use of NA as a remedy or management option that is based on the reliance that NA processes are able to achieve site-specific remediation or management objectives. Monitored natural attenuation should always include two essential aspects: source control and long-term performance monitoring.

It is a technique that has been used as site management and treatment technique for more than 30 years in the United States and only for a decade in Europe. The first articles on the subject appear at the end of the 1990s but the most significant publications appear in the early 2000s, which is consistent with the start of major European research projects on the subject (**Figure 1A**). Nevertheless, compared to other techniques, NA represents only an average of 5% of the techniques used for soil remediation. Phytoremediation and bioremediation represent 22% and 25% of techniques used using biological organisms whereas remediation (such as cleaning and excavation …) represents the main part with 44% (**Figure 1B**).

2.1 Role of natural attenuation in soil remediation

In 1999, natural attenuation is the "use of natural processes to avoid the spread of the contamination and reduce the amount of contaminants at contaminated sites," following the USEPA. It can be also defined by two terms (i) bioattenuation and (ii) intrinsic bioremediation. The term "bioremediation" is used when the contaminant sources have been removed by autochtonous microorganisms from the sites or if some hot spot contaminated is removed, but can also be used when the contaminant source is still present but less toxic. Thus, natural attenuation is widely accepted in the United States and in some European countries (Denmark, United Kingdom, Netherlands,

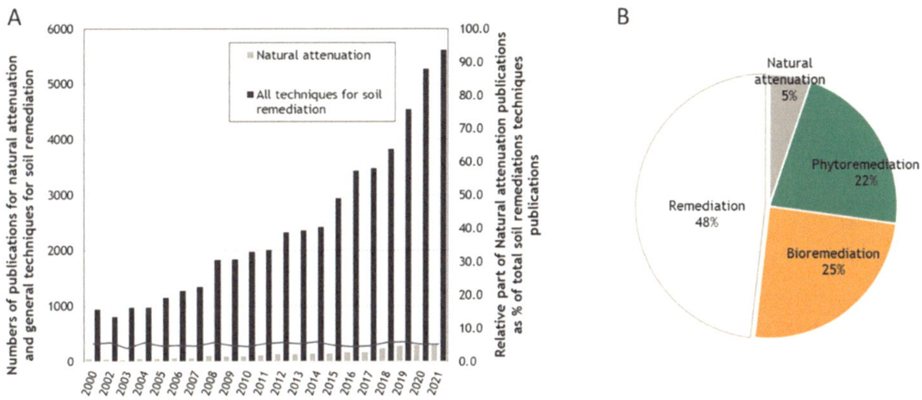

Figure 1.
A. Evolution of the number of annual scientific publications on general techniques of soil remediation (dark gray histogram) and natural attenuation (histogram light gray) in the international scientific literature over the period 2000–2021. The evolution of the relative share of publications on natural attenuation is represented by the black curve, which is estimated as % of the total number of publications on general techniques of soil remediation. B. Proportions of each technique during soil remediation over the period 2000–2021. Bibliometrics on the state of scientific and technological knowledge on natural attenuation technique has been evaluated with two search engines: Web of science and Medline, using these keywords "natural attenuation," "remediation," "phyroremediation," and "bioremediation" with different combinations.

Belgium, Germany ...) as a management option for contaminated soil [4]. Indeed, soils have a healing capacity with negative feedback mechanisms permitting the reinstallation of an optimal equilibrium in element cycling and an acceptable habitat for the biota. In addition, high diversity and flexibility of the biota are required to ensure the adaptation to actual conditions. Indeed, the biota has a constant lack of nutrients thus even contaminants may serve as a nutrient supply for soil microorganisms. As a consequence, contaminants enter into the biogeochemical element cycling and may disappear or accumulate in the environment. All these phenomena lead to a natural buffering and balancing system. Those natural processes are involved in environmental remediation, which can decrease the adverse impacts of contaminants and the environmental risk posed to the ecosystems. Natural attenuation used in environmental remediation is considered real technologies/processes occurring in natural reactors operating with or without engineering interventions. Depending on the scale of intervention, management activity is denominated differently, such as monitored natural attenuation, enhanced natural attenuation, soft remediation, close-to-nature bio- or ecotechnologies, *in situ* bioremediation, and phytoremediation [5]. Thus, most of the processes involved are common with the other traditional methods of remediation, such as sorption, dissolution, and biodegradation, which occur during phytoremediation or bioremediation. All the processes involved in natural attenuation are summarized in **Figure 2**.

2.2 Processes implied in natural attenuation

NA processes may be classified as either nondestructive processes, which do not destroy the contaminant but transfer it to another compartment or attenuate its concentration within a compartment, or destructive processes, which destroy contaminants *via* chemical or biological processes. This classification depends mainly on the type of contaminants. The main contaminants involved in the natural attenuation

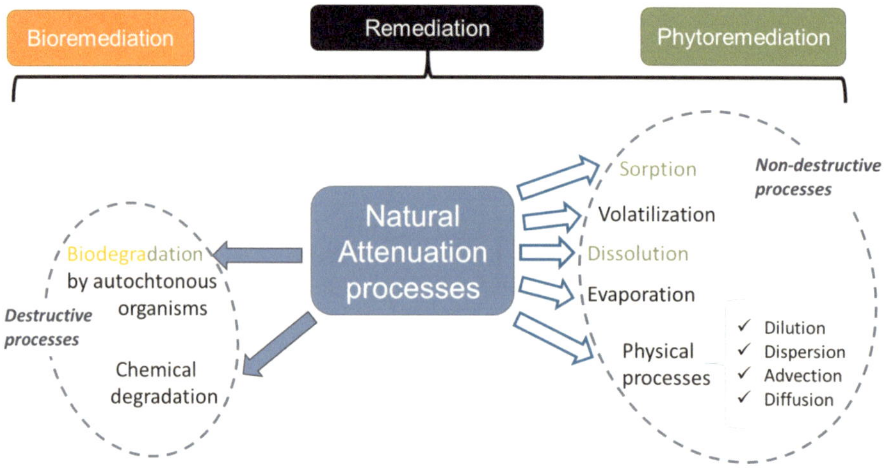

Figure 2.
Processes involved during natural attenuation and their similitude with the other common remediation techniques.

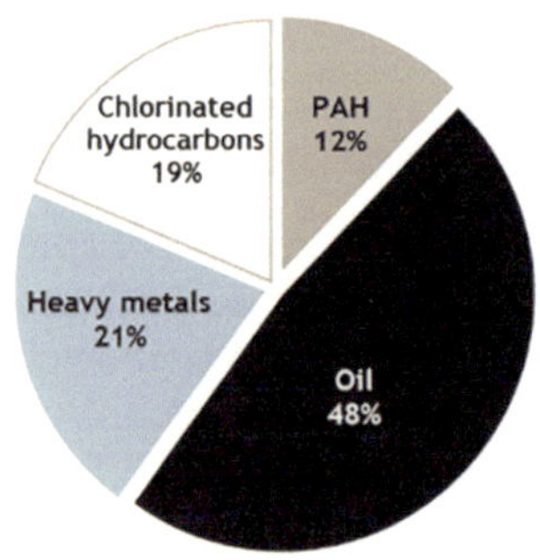

Figure 3.
Share of each pollutant treated in natural attenuation processes.

processes studied and published over the last 20 years are oils at 48%, polycyclic aromatic compounds (PAHs) at 12%, chlorinated hydrocarbons (CHC) at 19%, and heavy metals (metals and metalloids) at 21% (**Figure 3**). Oils are the primary contaminants of groundwater, while PAH, CHC, and heavy meals are the primary contaminants of soils.

Chlorinated hydrocarbon (CHC) is a generic term given to compounds containing chlorine, carbon, and hydrogen. The term can be used to describe organochlorine pesticides, such as lindane and DDT, industrial chemicals, such as polychlorinated biphenyls (PCB), and chlorine waste products, such as dioxins and furans. CHC is among the most common soil and groundwater contaminants due to their widespread use as dry-cleaning solvents and degreasing agents. These compounds are persistent in the environment and most bioaccumulates in the food chain [6]. PAHs are organic molecules with several benzene groups. Their origin in the environment can be natural (volcanic activity, formation of fossil fuels) or anthropic (industrial coal, petrochemical activity, and domestic activities). Metals are simple bodies that tend to lose electrons and differ in this respect from nonmetals. A metalloid is an intermediate element between a metal and a nonmetal. The major trace metals, because of their potential toxicity, are cadmium (Cd), cobalt (Co), chromium (Cr), copper (Cu), nickel (Ni), lead (Pb), and zinc (Zn). Metals are primarily present as cations in soil water. Arsenic is the main metalloid that poses environmental problems origin of trace metals and metalloids in the environment can be natural through the weathering of rocks which leads to their release, and/or anthropogenic. Industrial activities are the main origin of soil pollution by metallic components. Thus, the fate and transport of the pollutant are controlled by the physicochemical properties of the compound itself (such as its solubility in water, its hydrophobicity, its Henry K_h constant, or its water/organic carbon partition coefficient Koc), but also by the environment (pH, redox potential, granulometry, porosity, and organic carbon content) [7].

Natural attenuation processes can be divided into four categories:

- Nondestructive processes in which the mass of contaminants does not modify within a compartment (e.g., soil solution, groundwater, or the gaseous phase): advection (transport of a quantity of a given element by the movement of the surrounding environment), dispersion (=change of direction and velocity of a particle, often related to the movements of water), and molecular diffusion (substance migrates from an area of high concentration to an area of low concentration);

- Nondestructive processes in which the mass of contaminants will change within a compartment through the different phases of the transfer processes, such as dissolution, volatilization (=change of a substance into a gaseous state), and sorption (=property that some solid bodies have to retain the molecules of other substance on their surface);

- Destructive processes: that will mainly induce a reduction of contaminant mass and discharge through biodegradation and, to a much lesser extent, chemical degradation (photolysis, redox reactions, and chemical hydrolysis);

- Nondestructive phenomena grouped under the word "dilution," which does not refer to any physical process such as the processes described above and involves mixing of polluted and non-polluted water due to the action of various phenomena (e.g., recharge by rain).

The processes involved during natural attenuation for chlorinated hydrocarbons (CHC), polycyclic aromatic compounds (PAHs), metals, metalloids, and their main characteristics are summarized in **Table 1**. CHC and PAH can completely disappear from soil because of their high biodegradation but also form a by-product of degradation that could be potentially more toxic than the parent compound [4, 8]. Whereas metals and metalloids are contaminants that cannot be degraded and will be always persistent in soil but under another chemical form (speciation), less mobile and toxic, according to their surrounding environment.

2.3 How to monitor the processes of natural attenuation?

Natural attenuation relies on natural processes to reduce or "attenuate" contaminant concentrations in soil and groundwater to protect public health and safety, and the environment. Natural attenuation is sometimes mislabeled as a "do nothing" approach to site remediation. But the truth is that natural attenuation is seen as a "proactive approach" or "passive approach" in which the verification and monitoring of the processes of natural remediation will be more concerned rather than relying

	Chlorinated hydrocarbons	PAH	Metals and Metalloids
Example of contaminants	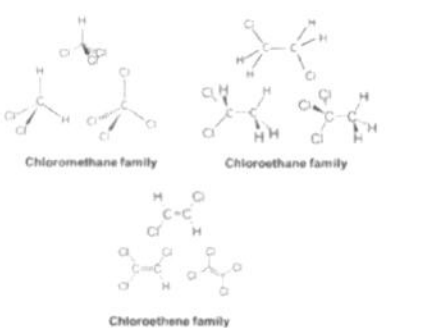	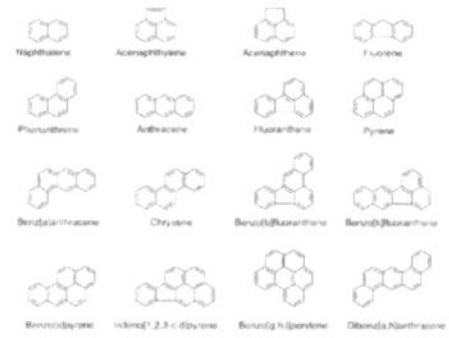	Cadmium (Cd), cobalt (Co), chromium (Cr), copper (Cu), nickel (Ni), lead (Pb), zinc (Zn), and arsenium (As)
Fate and properties	• Low solubility in water	• Low solubility in water	• High affinity of free ions for ligands
	• High volatility	• Low volatility except for naphthalene	• Mobility influenced by pH (acid pH favors solubilization and basic pH favors their sorption) and redox conditions
	• Low adsorption except on organic matter	• Hydrophobe	
	• Biodegradabilty significant	• High toxicity	• High toxicity depending on pH and redox conditions
		• Hight persistent	• High biodissolution
Processes involved	• Dispersion		• Dispersion
	• Volatilization	• Volatilization (only naphthalene)	• Volatilization (only Hg and Se species)
	• Sorption	• Adsorption on soild phase	• Adsorption/desorption depending on surrounding environment
			• Precipitation/dissolution depending on surrounding environment
			• Dilution
	• Abiotic degradation only for 1,1,1-Trichloroethane	• Photolysis	• Neither degradation nor biodegradation
	• Anaerobic biodegradation		
	• Aerobic biodegradation only for 1,1,1 Trichloroethane and Tetrachloroethylene	• High aerobic biodegradation for PAH with a maximum of five benzenic cycles	
Results	• Complete disappearance	• Complete disappearance	• No disappearance
	• Formation of degradation products that are potentially more toxic	• Formation of degradation products that are potentially more toxic	• Decrease toxicity and immobility

Table 1.
Main fate, properties, and processes involved for chlorinated hydrocarbons, PAHs (polycyclic aromatic compounds), metals, and metalloids during natural attenuation in soils.

totally on technical processes. Before natural attenuation should be proposed for any site, the contaminant concentrations must be evaluated to document that natural attenuation is occurring by scientists or operators. It is for this reason that monitored natural attenuation (MNA) refers to the use of NA as a remedy or management option. In general, monitoring involves taking soil and groundwater samples for analysis to determine the presence of contaminants, to check for further dispersion in contaminated areas, and to assess whether there is an improvement in soil functionality. In the case of soil contaminated with organic compounds, MNA is most effective where natural processes permanently degrade or destroy the contaminants. For inorganics pollutants, MNA is most effective where immobilization or radioactive decay is demonstrated to be occurring. Many reports [9–11] have been realized concerning the monitoring of MNA in groundwater, but very few works have been dedicated to the soil. According to this literature, some parameters can be proposed for MNA in soil according to the type of pollutants:

Organics:

- Concentration and distribution of contaminants, degradation products;
- Soil geochemical parameters (e.g., electron acceptors including nitrate, sulfate, and iron (III); chloride, pH, alkalinity, oxidation–reduction potential);
- Microcosm studies demonstrating the realization of biodegradation;
- Stable isotopic tools to monitor the degradation of contaminants;
- Mass flux calculations.

Inorganics:

- Concentration and distribution of contaminants;
- Soil geochemical parameters (pH, organic carbon content, ionic exchangeable capacity, granulometry, active lime …) to determine contaminant sorption and chemical speciation;
- Data to determine the contaminant immobilization and long-term stability;
- Mass flux calculations.

Finally, experts in sciences have pointed out several indicators or lines of evidence that can be used to prove that natural attenuation processes are operating at a site and are reducing the concentrations of contaminants [4]. The first indicator is historical trends showing a reduction in contaminant concentrations at the site over time and a plume that is either stable or declining. A stable or declining plume indicates that natural attenuation processes are working well in removing dissolve contaminants from soil or groundwater; and that the amount of contaminants removed is equal to or greater than the amount of contaminants added to the plume. The second indicator is the changes in chemical parameters of soil or groundwater. The processes of natural attenuation are directly related to changes in physical-chemical parameters of the environment, such as the level of oxygen, nitrate, sulfate, and other elements because of biological consumption; and the production of by-products, such as methane. These chemical indicators can be used to evaluate the site-specific potential for contaminants to be eliminated through the processes of natural attenuation. However, if one of the first two lines of evidence does not conclude that there is natural attenuation at the site, laboratory "microcosm" tests can be set up under controlled conditions and study of living actors at the site, including plants, fauna, and in particular native microorganisms, that potentially have the capacity to degrade and absorb contaminants. The rate of contaminant destruction by the various living actors can be measured and assessed directly.

2.4 Actors involved in natural attenuation

As mentioned above, natural attenuation is used to reduce the contaminant concentration in the environment through (i) biological processes, (ii) physical phenomena, and (iii) chemical reactions. Among them, biological processes, which are carried out by different actors, including microorganisms, animals, and plants, in close interaction with environmental factors, are the dominant mechanism for the cleanup of contaminated sites. Biological processes encompass two different ways (i) biodegradation/transformation mainly led by microorganisms and (ii) adsorption/uptake done by primarily plants and animals.

2.4.1 Microorganisms

Microorganisms, including bacteria, fungi, protozoa, and algae, represent the high biomass in the ecosystems, such as soil, groundwater, and ocean [12]. Moreover, the diversity of microbial communities living on earth is high and enormous. In soil, with hundreds of thousands of taxa per gram of soil, microbial diversity dominates soil biodiversity. Thank to high biomass and diversity, microorganisms are major players that contribute to the degradation/transformation and adsorption of contaminants. The processes by which microorganisms degrade or transform the contaminants are metabolic or enzymatic actions based on growth and cometabolism processes [13]. The rate of contaminant degradation depends on different parameters: (i) type and density of microorganisms, (ii) environmental factors, such as pH, presence or absence of O_2, nutrient concentrations, temperature, and moisture, and (iii) type and concentration of contaminants. For the first point on the type of microorganisms, bacteria and fungi are known to be the most studied and the most important actors in the degradation of contaminants. There are many reports on the degradation of contaminants by different type of bacteria and fungi (**Table 2**). Yeast, algae, and protozoa are also involved in biodegradation processes but there is not much research on their involvement. Moreover, the potential of contaminant degradation by mixed microbial communities that have high genetic information is higher than by a single bacterium [27]. Numerous environmental parameters are known to be drivers determining the rate of contaminant degradation by microorganisms because the growth of microorganisms depends on these environmental parameters. Particularly, under the presence (aerobic) or absence of O_2 (anaerobic), the efficiency of biodegradation by microbial communities is different from one contaminant to another. In some cases, combined aerobic/anaerobic conditions with a consortium of microbial communities

Contaminant types		Source of pollution	Mechanisms	Microorganisms involved	References
Organic compounds	**Polyaromatic hydrocarbons (PAHs)**	industrial pollution	metabolic/ enzyme	*Pseudomonas fluorescens* *Brevibacillus* sp., *Acinobacteria* *Mycobacterium*, *Corynebacterium*, *Aeromonas*, *Rhodococcus*, *Bacillus*	[14–18]
	Chlorinated hydrocarbons	industrial pollution	metabolic/ enzyme	*Pseudomonas* sp., *Burkholderia* sp., *Ralstonia*, *Achromobacter*, *Sphingomonas*, *Comamonas*, *Rhodococcus*, *Janibacter*, *Bacillus*, *Paenibacillus*, *Microbacterium*	[19–21]
inorganic compounds	**Arsenic**	natural sources, mining and smelting, industrial processes, agriculture practices	metabolic/ enzyme	*Sulforospirillum barnesii* *Sphaerotillus Leptothrix* group *Acidithiobacillus ferrooxidans*	[22, 23]
	Lead	historic mine sites	metabolic/ enzyme	*Cladosporium* sp. *Streptomyces ederensis* *Phyllobacterium loti* *Streptomyces fulvissimus*	[24]
	Chromium (Cr)	industrial pollution	biosorption	*Bacillus marisflavi*, *Arthrobacter* sp., *Sinorhizobium* sp. *Klebsiella* sp. *Streptomyces werraensis*	[25, 26]

Table 2.
Mechanisms and microorganisms involved for chlorinated hydrocarbons, PAHs (polycyclic aromatic compounds), metals, and metalloids during natural attenuation in soils.

are applied to get better results. In general, the degradation of contaminants is carried out by the action of a cocktail enzyme that is produced by divers of microorganisms [28]. It is known that the enzyme is active in the special condition of the soil, particularly pH soil. Therefore, biodegradation of contaminants depends well on pH soil. In general, microbial degradation occurs under a wide range of pH, but high rates of contaminant degradations were found under a pH of 6.5–8.5 in most contaminated sites [29]. Contaminants are divided into two subgroups: organic and inorganic contaminants. For organic contaminants, degradative microorganisms can completely degrade organic contaminants to CO_2 or transform them into smaller compounds. Microorganisms can use organic contaminants as a sole source of carbon and energy. This mechanism results in a complete degradation, called mineralization of organic contaminants [30, 31]. On the other hand, organic contaminants are transformed into a metabolite that is less toxic. Unlike organic contaminants, the metals, inorganic compounds, cannot be degraded but may be transformed into different forms which are nontoxic or immobile or are physically available for uptake by plants or animals.

2.4.2 Plants

Besides microorganisms, plants play also an important role in natural attenuation processes. In the contaminated sites, plants can develop different strategies in order to survive and thrive; therefore, plants must principally exhibit: (i) fast growth, (ii) high tolerance to one or cocktail of contaminants, and (iii) the ability to produce and secrete a different enzyme to assimilate, transform, metabolize, detoxify, and degrade various contaminants. The primary role of plants in natural attenuation processes is to uptake contaminants from soil/water, transport them, and accumulate them in their aerial parts. Unlike microorganisms which are not easily separated from soils, the plants after having absorbed and accumulated the contaminants in plant tissues can be removed from soils and thus reduce contaminant concentrations at the contaminated sites. The plants that have extraordinary ability to uptake, transport, and accumulate contaminants in their aerial parts are called hyperaccumulators. A potential and good candidate of hyperaccumulator for adsorption of contaminants has a ratio of bioaccumulation factor and translocation factor higher than 1. It is known that the ability to uptake and accumulation of contaminants in different parts of the plants depends on (i) the density and diversity of plants, (ii) the physicochemical properties of the soil, (iii) the weather and environmental conditions, and (iv) the nature of the contaminant itself [32].

Plant species within the same genus and different varieties of the same species may exert different patterns of contaminant uptake. Moreover, the concentration of contaminants accumulated in different parts of the plants, including roots, stems, and leaves, are different. Few works also showed that a rich diversity of plants enhance the natural attenuation in the case of polycyclic aromatic compounds (PAHs) [33].

With regard to soil properties, numerous studies highlight the role of soil texture (*i.e.,* sandy, sandy clay, loamy, and sandy loam), soil chemical characteristics (*i.e.,* pH, organic matter concentration, mineral concentration, cation exchange capacity, water solubility, and O_2 presence) in the capacity of plants to uptake and accumulate contaminants. These critical factors determine the nature of contaminants in the soil, in particular bioavailability form for plants, and thus the potential for uptake by crop plants [34]. In addition, some works [35] showed that in aerated soils, plants grow better than in non-aerated ones due to high respiration and good root functionality in the rhizosphere zone, better uptake of water and nutrients, and therefore better contaminant adsorption. Moreover, meteorological factors, such as ambient temperature, humidity, and wind speed may affect the adsorption of contaminants by crop plants by shaping the rate of evapotranspiration and thus the uptake of water, nutrient, and contaminants.

2.4.3 Soil fauna

In soils, in general, soil fauna, which is characterized by large numbers of animal species, plays a central role in the ecological functions of soil ecosystems. They participate in (i) organic matter mineralization and dynamics; (ii) support and regulation of primary production; and (iii) regulation of formation and development of soil [36]. Soil fauna composes of three groups according to their size: (i) microfauna includes organisms smaller than 0.1 mm in size or diameter, such as protozoans, nematodes, and rotiphers; and (ii) mesofauna consist of organisms in size from 0.1 to 2 mm, such as Acari, Collembolan, Enchytraeida, and others; and (iii) macrofauna are classified as organisms in size from 2 to 20 mm, including earthworms, termites, and others [37]. Many works showed that the soil fauna is sensitive to environmental changes and pollution [38, 39]. In highly polluted soils, some organisms can be found because organisms have adapted to the presence of the contaminants. The impact of contaminants on soil fauna depends on numerous factors, including soil properties, the concentration of the availability of contaminants, the sensibility of species, and avoidance strategy [40]. In turn, soil fauna through their activities can influence the distribution and the availability of contaminants that can improve the process of natural attenuation. In particular, numerous species of earthworms, such as *Eisenia fetida*, *Lumbricus terrestris*, *Lumbricus rubellus,* and *Aporrectodea caliginosa,* can survive in soils polluted with metals and even accumulate heavy metals, including Cd, Pb, Cu, and Zn [41]. The activities of earthworms were shown to increase the availability and mobility of heavy metals (*i.e.,* Cd, Pb, and Hg) [42–46]. Apart from earthworms, very few data have been demonstrated on the application of other soil fauna in the processes of natural attenuation.

2.4.4 Interaction between plant-fauna-microbe-soil

As we have shown above, in contaminated soils, the main actors that play a dominant role in the natural attenuation process are the microorganisms *via* their activities involved in the degradation of contaminants, and the plants *via* their capacity to uptake contaminants. However, it is important that, under real conditions, often in the presence of a cocktail of contaminants, the complexity of the interaction between plant-fauna-microorganisms under the influence of soil physicochemical parameters must be taken into account. The interactions between plants-microorganisms-fauna that can have an impact on the natural attenuation have been proposed and presented in **Figure 4**. The positive influences of vegetation, including habitat formation and root exudation production on microbial communities, have been already demonstrated in numerous works [47, 48]. An astonishing number of microbes are able to colonize plant

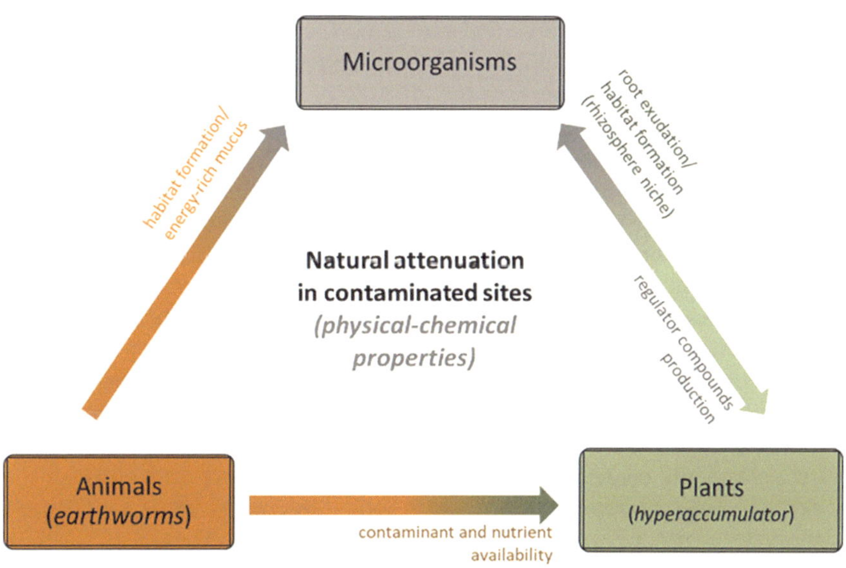

Figure 4.
The hypothesis of synergic relationships between soil-plants-animals-microorganisms in the process of natural attenuation at the contaminated sites.

roots very efficiency, forming the rhizosphere niche. In rhizosphere niche, microorganisms are equipped with many mechanisms to overcome toxicity and contribute to bioremediation. The mechanisms are (i) bioaccumulation in which microorganisms take up and accumulate contaminants in their cells, (ii) biotransformation in which contaminants are transformed into other less toxic compounds, and (iii) biosorption in which contaminants bind with cation-binding proteins present in the cell wall of microorganisms [47]. Root exudates produced by plants are organic acids, sugars, hormones, amino acids, vitamins, enzymes, and other types of primary and secondary metabolisms, which act as a source of nutrients for microbial growth and development. In other cases, some root exudates also prevent the growth of harmful microorganisms in plants [49]. They also play a significant role in bacterial quorum sensing and biofilm formation [50]. These lead to improve and increase the potential of microbial remediation. The interaction between the plant and the microorganisms is high complete. As cited above, plants help microorganisms to grow and resist contaminant toxicity, and on the other hand, microorganisms have positive and negative impacts on the plant. In general, rhizosphere microbiome consists of mainly beneficial bacteria and fungi that can have an impact on plant development. In fact, plants in retour can benefit from the production of regulator compounds called siderophores through the activities of rhizosphere microbiome to resist the toxicity of contaminants and to better grow in contaminated soils. The beneficial effects of siderophores produced by rhizosphere microorganisms on plant growth have been demonstrated in various works [51–53]. For example, the microbial produced-phytohormones by plant growth-promoting rhizobacteria (PGPR) promote plant growth and modify root architecture with an increase in root surface area leading to an increase of nutrient and contaminant uptake [54]. According to recent research, the utilization of PGPR for remediation is becoming a good strategy thanks to its various abilities to detoxify and degrade toxins, as well as its significant effects on plant growth promotion. Numerous bacteria, such as *Pseudomonas* sp., *Burkholderia* sp., *Bacillus* sp., *Bacillus* sp., *Penicillium* sp., *Streptomyces* sp., and *Endophytic* sp. are identified as the rhizobacteria species responsible for stress tolerance to organic and inorganic (heavy metals) contaminants in crop plants [53, 55, 56]. In conclusion, when the plants grow better, they can uptake more contaminants and thus increase the efficacy of the natural attenuation process. Regarding the soil fauna, among them, earthworms are known for their capacity to survive in contaminated soils and have many contributions on microbial growth through their activities, such as habitat formation (drilosphere zone) and support nutritional resources *via* the excretion of mucus [57]. On the other hand, earthworms can improve the soil's physical and chemical properties and increase the soil fertility through amelioration of microbial activities. Furthermore, according to the results demonstrate in many works, earthworms increase the bioavailability of inorganic contaminants (heavy metals) in soils, which is a primordial factor controlling the success of phytoremediation. Readers can find more information on the mechanisms direct and indirect of earthworms effect on soil microorganisms and plants in the paper published by Jusselme et al. (2020, 44]. So, high microbial growth and development thanks to the contribution of soil fauna (earthworms) may enhance the efficacy of the natural attenuation process in which microorganisms contribute to the degradation/transformation and adsorption of contaminants. Moreover, the activities of earthworms (soil fauna) are known to enhance the availability of nutrients as well as contaminants for plant health and development and also contaminant uptake by the plant. This leads to an increase in the potential of plants for the remediation of contaminants.

2.5 Factors enhancing natural attenuation processes

Remediation by enhanced natural attenuation (RENA) refers to an enhancement that can be done in soils to increase the degradation or sequestration of contaminants beyond what occurs naturally without our intervention [58–60]. In general, enhancements must be used at one time in order to maintain their effectiveness which involves a favorable balance between mass flux and system attenuation capacity. Most of the works for RENA are focused on organic contaminants and groundwater. Nevertheless, according to this literature, it is possible to give some examples of enhancement and their effectiveness. Thus, enhancement can involve improving nutrient, aeration, and moisture content for an *in situ* bioremediation approach, which has recorded many successes toward pollutant removal [58] in order to increase the biodegradation of organic pollutants. Moreover, some studies [60] have shown that the addition of an amendment can stimulate a natural consortium of bacteria to improve the degradation rate more effectively than without it. Natural consortia of microbes with additional species can be also added to increase the overall degradation of contaminants. Finally, sorption is another attenuation process that is possible to enhance by augmenting the stable fraction of organic carbon. Some natural attenuation processes cannot be implied for enhancement, especially in soil. For example, diffusion and dispersion are processes that cannot be feasible to increase.

2.6 Application and advantage/inconvenient of natural attenuation

Some countries have gained in experiences these last decades, especially the United States. Some European countries, such as Denmark, Netherlands, Belgium, Germany, and Sweden, had also developed for the first time some programs at the beginning of the 2000s. The purpose of these programs was to develop the protocols and real applications of MNA [4, 61]. These works published a review of different MNA concepts in those countries and provide an updated overview of the differences and similarities between them, with special attention to the return on experience and legal frameworks.

Nevertheless, there are numerous guides and protocols with the number increasing each year. It should be noted that these organizations do not have a common methodological tool for site management, which may explain the multiplicity of these protocols. A protocol can be defined as a strategy plan, a methodology to be followed for the proper use of natural attenuation as a rehabilitation technique. In general, three steps are always identified in the identification of the natural attenuation potential at a contaminated site:

- Development of conceptual models of the site hydrogeology and biogeochemical reactions at the site;

- Analysis of the measurements made on the site in order to quantify the natural attenuation processes (spatial and temporal evolution of the contaminants and their footprints);

- Establishment of a long-term monitoring program to verify the expected results.

All these studies and experiments have highlighted several advantages using when using monitored natural attenuation, such as low costs, minimal disturbance, and

Advantages	Disadvantages
• Overall costs are lower than other techniques.	• Longer periods of time may be required to mitigate contamination.
• Minimal disturbance to the site operations.	• Natural attenuation is not appropriate where imminent site risks are present.
• Potential use next and below buildings.	• Despite predictions that the contaminants are stationary, some migration of contaminants may occur.
• No generation of remediation wastes.	• Institutional controls may be necessary to ensure long-term protectiveness.
• Reduced risk of human exposure to contaminants near the pollution source.	• More efforts may be required in order to gain public acceptance of natural attenuation.
• Complete destruction of organic contaminants through biodegradation processes.	• Intermediates of biodegradation may be more toxic than the original contaminants.
• Stabilization of metals and metalloids.	• Some inorganics contaminants can be immobilized, such as metals, but they will not be degraded.

Table 3.
Advantages and disadvantages of monitored natural attenuation.

efficiency. Nevertheless, there are several limitations [3, 8, 62, 63] such as the long time periods before to mitigate contamination, the zero risks do not exist, and a lack of some data to model this process. All these advantages and disadvantages are displayed in **Table 3**.

3. Soil remediation by natural attenuation: What social acceptability?

Social acceptability has become a new decision-making standard for many development projects [64]. This concept gives rise to terminological debates [65], however, it can be defined as "the result of a process by which the parties concerned construct together the minimum conditions to be put in place, so that a project integrates [...] in its natural and human environment, [...] agreed in full knowledge of the facts by all the parties in a climate of fairness between the various experts." It is a question of finding acceptable reasons to justify a collective action that appears to be the best solution, including the status quo. This approach, therefore, implies taking into account the perception of public opinion, which conditions the appropriation of the project [66]. While the European Commission has presented a strategy in favor of soils, of the restoration and protection of soils by 2050, (knowing that 70% of them are degraded) [67], lifting the invisibility of soil pollution appears to be a prerequisite before considering the social acceptability of remediation projects and techniques to be preferred. It will then be necessary to consider the conditions for sociopolitical, community, and economic acceptability [68].

3.1 Invisible soil pollution?

3.1.1 A lack of scientific and technical knowledge

The distinction between healthy and polluted soil is made on the basis of the level of acceptable risk to the populations of the environment; this is specifically standardized by each society [69]. Moreover, the issue of polluted soil entered very late (1980s), in the environmental codes of the various European countries [70], whose pollution of certain regions was obscured and "sacrificed" to industrial development [69]. The soil has recently been the subject of particular attention [70]. The latest estimates for the EU28 show 2.8 million polluted plots [67], yet only about a quarter of these sites are included in databases. BASIAS and BASOL in France listed between 300,000 and 400,000 contaminated sites in 2006 [71]. In addition to the lack of knowledge in this area, there is a lack of information about the presence of pollution and its exact nature. This invisibility allows the problems associated with the implementation of clean-up projects to be hidden at different levels of decision-making and responsibility (land management, evolution of the activities concerned, public opinion...). The polluter-pays principle at the heart of the crucial question of costs requires defining responsibilities for contamination that often goes back several decades. Finally, the management of soil pollution has been done, for a long time, in a silent way in an inter-society between regulators, polluters, and engineers [67]. Moreover, the legislation concerning polluted soils in the European Union remains very fragmented.

3.1.2 Making soil pollution invisible: A pragmatic approach?

Soil and water pollution affects many territories with an industrial past and raises issues that are as much health-related (exposure to risks), socioeconomic (inequalities in terms of this exposure to risks, precariousness of employment areas, etc.), as symbolic (stigmatizing and potentially repulsive image for the establishment of companies) and, of course, political (management of costs, application of the polluter pays principle, policy of reconversion of territories, etc.) [71]. The success of an industrial reconversion implies, first of all a landscape reconstruction that hides the traces of the past. The invisibilization of pollution can be considered as the decision to "deal with" it, failing to eliminate it. Thus, in the Swansea Valley in Wales, the choice was made to leave the polluted soil on site and cover it with an impermeable material [67]. Physically concealed and polluted soils are no longer considered a problem for the redevelopment of contaminated sites. These camouflage strategies also involve the revegetation of polluted sites in order to make them landscape-pleasing and therefore potentially attractive [ibid]. The economic reconversion of certain sites (hosting businesses, tourism development, heritage enhancement, etc.) can thus lead to an invisibilization of the memory of the risk and of the pollution deemed the most embarrassing [72]. The removal of polluted soils can contribute to another form of invisibilization. Thus, soils evacuated to the outskirts of Paris during the remediation of various neighborhoods in the nineteenth century have been reintegrated into the contemporary urban fabric for the implementation of development projects and green spaces [73]. In the Nantes region, waste rock from mining sites has been moved to be used as a fill for development operations (roads, community facilities, and buildings) [74]. In addition, some wastes are stored in geological layers and others are outsourced to African and Asian countries. The concealment is also of a semantic nature, particularly in public discourse [75, 76]. Local elected officials, like some companies, seek first to reassure populations about the environmental and health impacts of polluted soil. Faced with a lack of technical and financial means, but also with political, economic, and electoral stakes, they may opt to minimize the problem.

3.2 Levers for making pollution visible

3.2.1 Visibility led by NGOs and citizens' movements?

Public actors play a real role in the reporting of soil pollution [73, 77], however, NGOs and citizens' associations often constitute the levers of its visibility. The first national policies in Europe were envisaged after the denunciation by militant associations (Green Peace, Robin Hood, Legambiente) of serious industrial incidents (toxic landfill of Lekkerkerk in the Netherlands, Seveso accidents in 1976, Chernobyl and Schweizerhalle in 1986) [78]. Specifically involved in the field of the environment on a national or international scale, or created in a specific context on a local or national scale, they endeavor to publicize the risks associated with pollution, to inform and raise awareness among populations that are sometimes unaware of or refuse to admit the dangers they face. Thus, the NGO France Nature Environment attempted in 2017, to mobilize the public around a European citizen's initiative called "L'Appel du sol" [67]. The actors who mobilize may come from intellectual professions with strong social and cultural capital, but previous participation in union organizations or protest movements may be a driving force, regardless of social category [75]. The visibility of pollution can be initiated by long-established residents who know the history of the site. However, the installation of new residents, in search of a pleasant living environment, encourages the denunciation of nuisances and the highlighting of an environmental heritage that is sometimes obscured by the older residents. The visibility of polluted soils corresponds to urban as well as rural contexts, land valuation, protection of spaces, promotion of new uses, and of sociological renewal (gentrification). When the "official" scientific studies are insufficient or instrumentalized, and/or the pollution is hidden, the groups involved in these actions set up their own expertise: field surveys, data collection, technical training, and collaboration with scientists. Their approach also involves publicizing their actions in order to put pressure on the public authorities and to enter into a more global framework of environmental activism by getting closer to other groups engaged in similar approaches.

3.2.2 Resilience and environmental justice: Mobilizing concepts

Originating in ecology, adapted to psychology, and then mobilized by the social sciences [67], the concept of resilience has become a dominant reference for territorial action. The term "territorial resilience" translates the "ability of certain territories to generate within themselves, capacities of resistance and adaptation to change, thus allowing them to maintain or recover the bases of their development and their specificity in the face of more or less brutal shocks" [79]. This is a concept used in cindynics, associated with abrupt events and their management. Soil pollution invites us to question the resilience of the sites concerned over longer timeframes and to question a latent risk. Moreover, the virtuous character attributed to the concept of territorial resilience must be put into perspective. Thus, territorial resilience operations such as the one carried out in the Swansea Valley previously mentioned [77], do not imply the elimination of pollution and associated risks. Moreover, the mobilization of this concept of resilience tends to polarize the debates on the only local capacities of adaptation to socioeconomic and environmental changes without questioning the structural causes; this annihilates any reflection on the issues of territorial inequalities and environmental justice. Resilience is then seen as an incentive for local actors to adapt by their own means in a context of state disengagement.

Not all territories are equal in their capacity to adapt. Resilience operations favor potentially attractive territories, where the cost of soil decontamination can be assumed thanks to high profitability of land and real estate. These issues have been raised by research on the concept of environmental justice [80]. They highlight the link between exposure to a degraded environment and belonging to disadvantaged social categories [81] and ethno-racial minorities. The case of the city of New Haven studied and mapped [82] is edifying in this respect. More generally, the transfer of waste from the centers to their margins or from urban spaces to agricultural spaces is indicative of this environmental injustice, whose denunciation initiated in the 1970s in the United States [83] does not stop mobilizing.

3.3 Acceptability of the clean-up process and technical choices

3.3.1 Soil remediation through the lens of ecosystem services

The acceptability of soil remediation projects depends on the benefits known to society and the value attributed to them. The social benefits can therefore be considered through the prism of ecosystem services. This concept has been mobilized and widely publicized by the Millennium Ecosystem Assessment (2015); the objective was to promote the protection of ecosystems by assigning economic and social value to the services provided by them [84]. Ecosystem services can therefore be defined as the benefits provided by ecosystems to human societies. A distinction is made between production (or supply) services, regulation services, and cultural services [84]. Despite the reservations which are made by ecologists and sociologists among others with regard to this concept and the reflections as to a “commodification of nature” [85], this can be useful here to consider the potential economic and social benefits of the soil remediation, particularly with natural attenuation. The most directly perceptible benefit for the population is undoubtedly landscape and esthetic. The decontamination of polluted sites can on radically modify the urban landscape and the image of districts or cities sometimes stigmatized by their industrial or mining past, and thus procure an embellishment to which the local populations are sensitive [65]. The revegetation of these soils can provide spaces for relaxation and leisure. In this sense, these are the benefits associated with cultural services that can be highlighted. Soils also provide regulatory services. In the context of sustainable city projects, soil remediation can contribute to the objectives of reducing greenhouse gases and improving air quality. Soils and plants store carbon in their tissues via photosynthesis, and participate in the maintenance or dissemination of a certain diversity of biodiversity, (biodiversity of soils is too often unknown), and can be integrated into larger projects for the maintenance or development of biodiversity. The greater permeability of these soils is an asset to limit the runoff and potential flooding in certain cases and restoration of the water cycle more generally, including filtering and purification functions, provided by soils and vegetation [44]. Soils also provide production services; they are the basis of agricultural activity and contribute to product quality. The assessment of these social and environmental amenities provided by soils remediation projects are, however, for the most part, complex to assess and account economically, in particular regulation and cultural services. The monetary calculation of the direct or indirect services rendered could however minimize the costs of soil rehabilitation projects and facilitate their wider implementation. In a context of scarcity and overbidding of land, the remediation, restoration, and then requalification of these soils is a real challenge [71]. The SES approach implies choices regarding the policy to be implemented. Should the multifunctionality of soils be preserved whatever the current use of the land? This is the preferred option in the Netherlands. Should the soil functions to be protected and is identified according to the current land use and its possible use? Canada and the Flemish region in Belgium are moving in this direction [Nowak 2003].

3.3.2 The acceptability of the techniques chosen: A question of perception and interpretation by the actors concerned

The perception of these remediation projects, by the population concerned is influenced by multiple factors: first, the identification of the risk associated with soil pollution and the potential benefits expected from natural attenuation. This identification is closely linked to knowledge of the health risks involved. It was highlighted in a Quebec mining site, that the knowledge by all of strong soil pollution, whose effects on the health of populations are clearly highlighted, facilitates the acceptance of soil remediation projects. In this case, the benefit is clearly identifiable, so the populations are extremely favorable to a method of depollution considered as ecological [86]. However, it is not necessarily the case in urban areas where pollution is old and associated with activities considered to be less polluting. Thus, the spreading of Parisian mud on the fields of farmers located in the immediate suburbs of Paris in the nineteenth century was not initially considered as a polluting activity [87]. In addition, the renewal of the population in a good number of urban regions leads to a lack of knowledge of the history of soils and associated pollution. In most cases, the esthetic and landscaping criteria have an essential role in the reception that can be given to this type of project [88]. The remediation of soils in neighborhoods where the image is devalued by an industrial or mining past constitutes a benefit clearly identifiable by the population who have been living there for a long time or more recently. Revegetation of the soils is particularly often equated with embellishment and an improvement of the living environment from an ecological point of view. Consultation on the landscapes desired by local residents requires a time of information and consultation that is added to the time necessary to obtain the first effects of the different soil remediation methods. In this regard, it should be emphasized that local communities such as companies specializing in soil treatment are often ill-informed and poorly trained or little trained in this type of alternative techniques and prefer to apply better known and better-controlled methods, such as excavation and backfilling of polluted areas. It seems that alternative remediation, particularly natural attenuation, is struggling to get out of the purely scientific and experimental sphere. In terms of costs, *in situ* alternative natural treatment is a much less expensive technique than conventional techniques, however, it still seems to be little applied [89]. The low use of these treatments can be explained by a lack of knowledge, lack of experience, uncertainty about treatment effectiveness, uncertainty about cleaning performance, and the longer time usually needed for *in situ* remediations. The time required to obtain significant results is a constraint both for development companies, local authorities, and for the population. In the Netherlands, the tolerated duration of soil treatment by AN is about 30 years; in the United States, the duration beyond which another technical choice will be preferred is not fixed; it must be “reasonable” [7]. The results must be perceptible to the residents, who must be reassured by precise information on the progress of the treatment phase. In the process of acceptability of natural attenuation, an articulation between these different temporalities constitutes an issue to be taken up.

4. Conclusions

Natural attenuation and monitored natural attenuation (MNA) use the natural capabilities of microorganisms in close interaction with plants and fauna at the site to degrade contaminants as well as the geochemical immobilization and NA processes in the environment. MNA can be an efficient, green, and sustainable approach for contaminated soil remediation when the source of contamination is controlled and when potential risks for human and ecological systems can be managed in an acceptable manner. In general, MNA alone may be unable to reach cleanup goals within a reasonable timeframe, therefore MNA is often associated with other remediation methods. Its use should be most favored within vulnerable habitats where the implementation of other remedial activities might cause additional unacceptable environmental damages.

Conflict of interest

The authors declare no conflict of interest.

Author details

Clarisse Balland-Bolou-Bi[1], Florence Brondeau[2] and My Dung Jusselme[1*]

1 Laboratoire Eau Environnement et Systèmes Urbains (Leesu), Université Paris, Créteil, France

2 Laboratoire Espaces, Nature et Culture (ER ENeC), Université Paris-Sorbonne, Paris, France

*Address all correspondence to: thi-my-dung.jusselme@u-pec.fr

References

[1] Eugenio NR, Montanarella L, McLaughlin M, Vargas R. Environment pollution journal – Special issue: Global status of soil pollution. Environmental Pollution. 2020;**115231**

[2] Duda AM. Addressing nonpoint sources of water pollution must become an international priority. Water Science and Technology. 1993;**28**(3–5):1-11. DOI: 10.2166/wst.1993.0398

[3] USEPA. Proceedings of the Symposium on Natural Attenuation of Chlorinated Organics in Ground Water, Report EPA/540/R-97/504. Dallas, Texas: USEPA; 1996

[4] Rügner H, Finkel M, Kaschl A, Bittens M. Application of monitored natural attenuation in contaminated land management—A review and recommended approach for Europe. Environmental Science & Policy. 2006; **9**(6):568-576

[5] Gruiz K. Natural attenuation in contaminated soil remediation. In: Press C, editor. Engineering Tools for Environmental Risk Management. 1st ed. London: CRC Press; 2019. p. 107

[6] Glossary EEA. http://balkans.unep.ch/glossary/gloassaryb.html. Available from: http://balkans.unep.ch/glossary/gloassaryb.html

[7] Nowak C, Mossmann JR, Saada A. Etat des connaissances sur l'atténuation naturelle: mécanismes et mise en œuvre. Vol. 97. Paris: Rapport BRGM/RP-51960-FR; 2003. p. 9

[8] Morgan P, Sinke AJC. Monitored natural attenuation. L Contam Reclam. 2005;**13**(3):247-265

[9] NJDEP, New Jersey Department of Environmental Protection. Site Remediation Program: Monitored Natural Attenuation Technical Guidance. New Jersey: NJDEP; 2012

[10] Wiedemeier TH, Barden MJ, Dickson WZ, Major D. Multiple lines of evidence used to evaluate natural attenuation and enhanced remediation. In: Practical Handbook of Environmental Site Characterization and Groundwater Monitoring. 2nd ed. London: CRC Press; 2005

[11] Wilson JT. An Approach for Evaluating the Progress of Natural Attenuation in Groundwater. National Risk Management Research Laboratory: US EPA; 2011

[12] Álvarez-Rogel J, Peñalver-Alcalá A, Jiménez-Cárceles FJ, Carmen Tercero M, Nazaret G-AM. Evidence supporting the value of spontaneous vegetation for phytomanagement of soil ecosystem functions in abandoned metal(loid) mine tailings. Catena. 2021;**201**:105191

[13] Rozenbaum O, Barbanson L, Muller F, Bruand A. Significance of a combined approach for replacement stones in the heritage buildings' conservation frame. Comptes Rendus - Geosci. 2008;**340**(6):345-355

[14] Abbasnezhad H, Gray M, Foght JM. Influence of adhesion on aerobic biodegradation and bioremediation of liquid hydrocarbons. Applied Microbiology and Biotechnology. 2011; **92**(4):653-675

[15] Tyagi M, da Fonseca MMR, de Carvalho CCCR. Bioaugmentation and biostimulation strategies to improve the effectiveness of bioremediation processes. Biodegradation. 2011;**22**(2): 231-241

[16] Brezna B, Khan AA, Cerniglia CE. Molecular characterization of dioxygenases from polycyclic aromatic hydrocarbon-degrading mycobacterium spp. FEMS Microbiology Letters. 2003; **223**(2):177-183

[17] Khan AA, Wang RF, Cao WW, Doerge DR, Wennerstrom D, Cerniglia CE. Molecular cloning, nucleotide sequence, and expression of genes encoding a polycyclic aromatic ring dioxygenase from mycobacterium sp. strain PYR-1. Applied and Environmental Microbiology. 2001; **67**(8):3577-3585

[18] Gkorezis P, Daghio M, Franzetti A, Van Hamme JD, Sillen W, Vangronsveld J. The interaction between plants and bacteria in the remediation of petroleum hydrocarbons: An environmental perspective. Frontiers in Microbiology. 2016;7:1836

[19] Petrić I, Hršak D, Fingler S, Vončina E, Ćetković H, Begonja Kolar A, et al. Enrichment and characterization of PCB-degrading bacteria as potential seed cultures for bioremediation of contaminated soil. Food Technology and Biotechnology. 2007;**45**:11-20

[20] Taguchi K, Motoyama M, Iida T, Kudo T. Polychlorinated biphenyl/biphenyl degrading gene clusters in Rhodococcus sp. K37, HA99, and TA431 are different from well-known bph gene clusters of Rhodococci. Bioscience, Biotechnology, and Biochemistry. 2007; **71**(5):1136-1144

[21] Kanade SN, Ade AB, Khilare VC. Malathion degradation by Azospirillum lipoferum Beijerinck. Journal of Scientific Research and Report. 2012; **2**(1):94-103

[22] Fukushi K, Sasaki M, Sato T, Yanase N, Amano H, Ikeda H. A natural attenuation of arsenic in drainage from an abandoned arsenic mine dump. Applied Geochemistry. 2003;**18**(8): 1267-1278

[23] Wang G, Su MY, Chen YH, Lin FF, Luo D, Gao SF. Transfer characteristics of cadmium and lead from soil to the edible parts of six vegetable species in southeastern China. Environmental Pollution. 2006;**144**:127-135

[24] Newsome L, Bacon CGD, Song H, Luo Y, Sherman DM, Lloyd JR. Natural attenuation of lead by microbial manganese oxides in a karst aquifer. Science of the Total Environment. 2021; **754**:142312

[25] Jobby R, Jha P, Gupta A, Gupte A, Desai N. Biotransformation of chromium by root nodule bacteria Sinorhizobium sp. SAR1. PLoS One. 2019;**14**(7): e0219387

[26] Latha S, Vinothini G, Dhanasekaran D. Chromium [Cr(VI)] biosorption property of the newly isolated actinobacterial probiont Streptomyces werraensis LD22. 3. Biotech. 2015;**5**(4):423-432. DOI: 10.1007/s13205-014-0237-6

[27] Fritsche W, Hofrichter M. Aerobic degradation of recalcitrant organic compounds by microorganisms, in environmental biotechnology. In: Jördening H-J, Winter J, editors. Concepts and Applications. Weinheim, FRG: Wiley-VCH Verlag GmbH & Co. KGaA; 2005

[28] Zhang T, Zhang H. Microbial consortia are needed to degrade soil pollutants. Microorganisms. 2023;**10**

[29] Cases I, de Lorenzo V. Genetically modified organisms for the environment: Stories of success and failure and what we have learned from

them. International Microbiology. 2005; **8**(3):213-222

[30] Sørensen SR, Holtze MS, Simonsen A, Aamand J. Degradation and mineralization of nanomolar concentrations of the herbicide dichlobenil and its persistent metabolite 2,6-dichlorobenzamide by Aminobacter spp. isolated from dichlobenil-treated soils. Applied and Environmental Microbiology. 2007;**73**(2):399-406

[31] Holtze MS, Sørensen SR, Sørensen J, Aamand J. Microbial degradation of the benzonitrile herbicides dichlobenil, bromoxynil and ioxynil in soil and subsurface environments–insights into degradation pathways, persistent metabolites and involved degrader organisms. Environmental Pollution. 2008;**154**(2):155-168

[32] Christou A, Papadavid G, Dalias P, Fotopoulos V, Michael C, Bayona JM, et al. Ranking of crop plants according to their potential to uptake and accumulate contaminants of emerging concern. Environmental Research. 2019;**170**: 422-432

[33] Bandowe BAM, Leimer S, Meusel H, Velescu A, Dassen S, Eisenhauer N, et al. Plant diversity enhances the natural attenuation of polycyclic aromatic compounds (PAHs and oxygenated PAHs) in grassland soils. Soil Biology and Biochemistry. 2019;**129**: 60-70

[34] Vasudevan D, Bruland GL, Torrance BS, Upchurch VG, MacKay AA. pH-dependent ciprofloxacin sorption to soils: Interaction mechanisms and soil factors influencing sorption. Geoderma. 2009;**151**(3):68-76

[35] Juraske R, Castells F, Vijay A, Muñoz P, Antón A. Uptake and persistence of pesticides in plants: Measurements and model estimates for imidacloprid after foliar and soil application. Journal of Hazardous Materials. 2009;**165**(1–3):683-689

[36] Briones MJI. The serendipitous value of soil Fauna in ecosystem functioning: The unexplained explained. Front Environmental Sciences. 2018;**6**:149

[37] DeLuca TH, Pingree MRA, Gao S. Chapter 16-assessing soil biological health in forest soils. In: Busse M, Giardina CP, Morris DM, Page-Dumroese DSBT-D in SS, editors. Global Change and Forest Soils. Amsterdam: Elsevier; 2019. pp. 397-426

[38] Copley J. Ecology goes underground. Nature. 2000;**406**(6795):452-454. DOI: 10.1038/35020131

[39] Dick RP. A review: Long-term effects of agricultural systems on soil biochemical and microbial parameters. Agriculture, Ecosystems and Environment. 1992;**40**(1):25-36

[40] Gillet S, Ponge JF. Humus forms and metal pollution in soil. European Journal of Soil Science. 2002;**53**(4):529-540. DOI: 10.1046/j.1365-2389.2002.00479.x

[41] Morgan JE, Morgan AJ. The accumulation of metals (Cd, Cu, Pb, Zn and Ca) by two ecologically contrasting earthworm species (Lumbricus rubellus and Aporrectodea caliginosa): Implications for ecotoxicological testing. Applied Soil Ecology. 1999;**13**(1):9-20

[42] Blouin M, Hodson ME, Delgado EA, Baker G, Brussaard L, Butt KR, et al. A review of earthworm impact on soil function and ecosystem services. European Journal of Soil Science. 2013; **64**(2):161-182. DOI: 10.1111/ejss.12025

[43] Sizmur T, Hodson ME. Do earthworms impact metal mobility and

availability in soil? – A review. Environmental Pollution. 2009;**157**(7): 1981-1989

[44] Jusselme MD. In: Bousserrhine N, editor. Phytoremediation: An Ecological Solution for Decontamination of Polluted Urban Soils. Rijeka: London, UK.IntechOpen; 2020. p. Ch. 11. DOI: 10.5772/intechopen.93621

[45] Jusselme MD, Poly F, Miambi E, Mora P, Blouin M, Pando A, et al. Effect of earthworms on plant Lantana camara Pb-uptake and on bacterial communities in root-adhering soil. Science of Total Environment. 2012;**416**: 200-207

[46] Jusselme MD, Miambi E, Mora P, Diouf M, Rouland-Lefèvre C. Increased lead availability and enzyme activities in root-adhering soil of Lantana camara during phytoextraction in the presence of earthworms. Science of the Total Environment. 2013;**445–446**:101-109

[47] Saeed Q, Xiukang W, Haider FU, Kučerik J, Mumtaz MZ, Holatko J, et al. Rhizosphere bacteria in plant growth promotion, biocontrol, and bioremediation of contaminated sites: A comprehensive review of effects and mechanisms. International Journal of Molecular Sciences. 2021;**22**(19):10529

[48] Patra DK, Pradhan C, Patra HK. Toxic metal decontamination by phytoremediation approach: Concept, challenges, opportunities and future perspectives. Environmental Technology and Innovation. 2020;**18**:100672

[49] Khan N, Ali S, Shahid MA, Mustafa A, Sayyed RZ, Curá JA. Insights into the interactions among roots, rhizosphere, and Rhizobacteria for improving plant growth and tolerance to abiotic stresses: A review. Cell. 2021; **10**:1551

[50] Vishwakarma K, Kumar N, Shandilya C, Mohapatra S, Bhayana S, Varma A. Revisiting plant-microbe interactions and microbial consortia application for enhancing sustainable agriculture: A review. Frontiers in Microbiology. 2020;**11**:560406

[51] Dubeikovsky AN, Mordukhova EA, Kochetkov VV, Polikarpova FY, Boronin AM. Growth promotion of blackcurrant softwood cuttings by recombinant strain Pseudomonas fluorescens BSP53a synthesizing an increased amount of indole-3-acetic acid. Soil Biology and Biochemistry. 1993; **25**(9):1277-1281

[52] Ramírez M, Valderrama B, Arredondo-Peter R, Soberón M, Mora J, Hernández G. Rhizobium etli genetically engineered for the heterologous expression of Vitreoscilla sp. hemoglobin: Effects on free-living and Symbiosis. Molecular Plant-Microbe Interactions. 1999;**12**(11):1008-1015. DOI: 10.1094/MPMI.1999.12.11.1008

[53] Ke X, Feng S, Wang J, Lu W, Zhang W, Chen M, et al. Effect of inoculation with nitrogen-fixing bacterium pseudomonas stutzeri A1501 on maize plant growth and the microbiome indigenous to the rhizosphere. Systematic and Applied Microbiology. 2019;**42**(2):248-260

[54] Gou W, Tian L, Ruan Z, Zheng P, Chen F-J, Zhang L, et al. Accumulation of choline and glycinebetaine and drought stress tolerance induced in maize (Zea Mays) by three plant growth promoting rhizobacteria (PGPR) strains. Pakistan Journal of Botany. 2015;**47**: 581-586

[55] Awan SA, Ilyas N, Khan I, Raza MA, Rehman AU, Rizwan M, et al. Bacillus siamensis reduces cadmium accumulation and improves growth and

antioxidant defense system in two wheat (Triticum aestivum L.) varieties. Plants. 2020;**9**(7):878

[56] Akhtar N, Ilyas N, Yasmin H, Sayyed RZ, Hasnain Z, A Elsayed E, et al. Role of Bacillus cereus in improving the growth and Phytoextractability of Brassica nigra (L.) K. Koch in chromium contaminated soil. Molecules. 2021;**26**(6):1569

[57] Medina-Sauza RM, Álvarez-Jiménez M, Delhal A, Reverchon F, Blouin M, Guerrero-Analco JA, et al. Earthworms building up soil microbiota, a review. Front. Environmental Sciences. 2019;7(JUN):1-20

[58] Odukoya J, Lambert R. Remediation by enhanced natural attenuation (RENA): A beneficial strategy for polyaromatic hydrocarbon degradation and agrifood production. Proc World Congr New Technol (NewTech 2015). 2015;**183**:183

[59] Chikere CB, Azubuike CC, Fubara EM. Shift in microbial group during remediation by enhanced natural attenuation (RENA) of a crude oil-impacted soil: A case study of ikarama community, Bayelsa, Nigeria. 3 Biotech. 2017;7(2):1-11

[60] Early T, Borden B, Heitkamp M, Looney B, Major D, Waught J, et al. Permanent enhanced attenuation: A reference guide on approaches to increase the natural treatment capacity of a system. 2006

[61] Declercq I, Cappuyns V, Duclos Y. Monitored natural attenuation (MNA) of contaminated soils: State of the art in Europe—A critical evaluation. Science of the Total Environment. 2012;**426**:393-405

[62] McBride MB. Toxic metals in sewage sludge-amended soils: Has promotion of beneficial use discounted the risks? Advances in Environmental Research. 2003;**8**:5-19

[63] EPA U, Land and Emergency Management L, EPA. For Underground Storage Tank Sites a Guide for Corrective Action Plan Reviewers: Chapter V Landfarming. Epa 510-B-17-003. Washington, DC: EPA; 2017

[64] Gendron C, Yates S, Motulsky B. L'acceptabilité sociale, les décideurs publics et l'environnement: légitimité et défis du pouvoir exécutif. [Vertig O] La Rev électronique en Sci l'environnement. 2016;**16**(1):1-27

[65] Almaric M, Cirelli C, Larrue C. Quelle réception sociale pour l'ingénierie écologique industrielle ? L'insertion socio-territoriale des zones humides artificielles. VertigO-la Rev électronique en Sci l'environnement. 2015;**30**(15):1-13

[66] Saucier C, Côté G, Fortin M-J, Jean B, Danielle Lafontaine ÉF, Guillemette M, et al. Développement territorial et filière éolienne. Rimouski, Canada: Uqar; 2009. p. 216

[67] Morel, Journel C, Gay G, Ferrieux C. La résilience territoriale comme principe et comme volonté Réflexions à partir de la question de la pollution des sols dans des territoires (dés)industrialisés. VertigO - la Rev électronique en Sci l'environnement. 2018;**30**

[68] Wolsink M. Contested environmental policy infrastructure: Socio-political acceptance of renewable energy, water, and waste facilities. Environmental Impact Assessment Review. 2010;**30**(5):302-311

[69] KREIT Jean-François. Sols contaminés, sols à décontaminer. In: Presses universitaires de Saint Louis. In: Les notions de sols pollués et la

décontamination des sols sur le plan technique CEDRE, SERES, UCL. CÉDRE, CEN. Bruxelles: Presses de l'Université Saint-Louis; 1996. pp. 1-13

[70] Elsig A. La gestion des sols pollués en Europe (XX e - siècles). Paris: Publisher EHNE; 2018. p. 5

[71] Leyval C, Cébron A, Beguiristain T, Faure P, Ouvrard S. Pollution Mitigation: Natural Attenuation of Organic Pollutants. Stuttgart: Schweizerbart Science Publishers; 2017

[72] Zanetti T. La pollution des sols dans les territoires (post) industriels: la résilience entre norme institutionnelle et cadre de lutte socio-environnementale. Géocarrefour; 2018;**11744**:1-18

[73] Rémy E, Gitton C, Yoann V, Brondeau F, Charvet R, Nold F, et al. NoL'économie circulaire: cercle vertueux ou cercle vicieux? Le cas de l'utilisation de terres maraîchères pour aménager des espaces verts urbains. Géocarrefour; 2018;**92**(2):1-30

[74] Hadna S. Controverse autour des stériles uranifères: de la mise à l'agenda d'un problème public à la remise en cause de l'expertise. Géocarrefour; 2018; **11850**:1-30

[75] Mesini B. Le laboratoire cévenol de l'après-mine. Une coextensivité des causes et des responsabilités minières, environnementales et sanitaires. Géocarrefour. 2018;**11887**:1-26

[76] Hadna S. Controversy around uranium waste: From putting a public problem on the agenda to challenging expertise. Comparative study of two former uranium mines: La commanderie (Vendée/Deux-Sèvres) and pen ar ran (Loire-Atlantique). Géocarrefour. 2018; **92**(2):1-24

[77] Ferrieux C, Le Noan R. Make the valley green again: la gestion des sols pollués au cœur de la réhabilitation de la basse vallée de Swansea (Pays de Galles). Géocarrefour. 2018;**92**(2):1-26

[78] Bertrand M. La protection des sols dans le cadre de l'Union européenne. J. Moulin: Sciences de l'Homme et Société. Université Lyon 3; 2018

[79] Hamdouch A, Deprat M, Tanguy C. Mondialisation et résilience des territoires. Presse Quebec: de l'Université de Quebec; 2012

[80] Gobert J. Les inégalités environnementales: une problématique socio-spatiale multi-dimensionnelle Julie Gobert To cite this version: HAL Id : hal-01786129 Justices et Injustices environnementales Cyrille Harpet Philippe Billet Jean-Philippe Pierron L' Harmattan. Nanterre: Presse universitaire de Paris; 2018

[81] Larrère C. Les inégalités environnementales. Paris: Puf-Vie de; 2017. p. 104

[82] Blanchon D, Moreau S, Veyret Y. No title. Ann Georgr. 2009;**665–666**(1–2): 35-60

[83] Paddeu F. D' un mouvement à l' autre: des luttes contestataires de justice environnementale aux pratiques alternatives de justice alimentaire ? Flaminia Paddeu. Nanterre: Presse universitaire de Paris; 2017

[84] Méral P. Les services écosystémiques. Versailles: Quae

[85] Maris V. Nature à vendre. Les limites des services écosystémiques. Versailles: Publisher Quae; 2014. p. 92

[86] Vodouhe FG, Khasa DP. Local community perceptions of mine site

restoration using phytoremediation in Abitibi-Temiscamingue (Quebec). International Journal of Phytoremediation. 2015;**17**:962-972

[87] Chaline C, Barles S. La ville délétère, médecins et ingénieurs dans l'espace urbain, XVIIIe-XIXe siècle. Annales le Geographie. 1999;**608**:436

[88] Origo N, Wicherek S, Hotyat M. Réhabilitation des sites pollués par phytoremédiation. VertigO. 2012;**12**(2): 1-20

[89] Montpetit É, Lachapelle E. Information, values and expert decision-making: The case of soil decontamination. Policy Sciences. 2016; **48**:1-17